First, Break IT

How Application Networks are Changing the Role of the CIO

Edited by Ross Mason

MuleSoft, Inc.
77 Geary Street, Suite 400
San Francisco, CA 94108
info@mulesoft.com
www.mulesoft.com
Edited by Ross Mason
Co-edited by Shana Pearlman
Cover design by Christopher Payne

First published by Dog Ear Publishing
4011 Vincennes Rd
Indianapolis, IN 46268
www.dogearpublishing.net

ISBN: 978-1-4575-4685-3

This book is printed on acid-free paper.

Printed in the United States of America

Table of Contents

Foreword on the business urgency of innovation by Michael Capellas, Founder and CEO of Capellas Partners

Foreword

By Michael Capellas, Founder and CEO of Capellas Partners

We are constantly reminded in our daily lives that the rapid adoption of technology-led innovation is disrupting traditional business models in nearly every industry.

For example, retail has moved beyond online shopping to quick-response, mobile-capable commerce. Consumers benefit from broader product and service selections, nearly complete price transparency, and the availability of an array of delivery options. Unmanned drones will soon transport many consumer products for same-day delivery directly to the home. Financial markets have recognized the far-reaching effects as these new business models achieve unparalleled scale. Amazon now has nearly 300 million active customers globally and a market capitalization over $300 billion, exceeding Wal-Mart by nearly 50 percent. The Alibaba Group claims nearly 300 million active mobile users and its 2014 IPO was the largest ever recorded.

In transportation, Uber has utilized an array of digital technologies to dramatically redefine the customer experience and create a powerful global brand. By late 2014, Uber claimed more than one million rides per day. Traditional car manufacturers are not only facing competition from well-financed new entries like Tesla, whose market value is now more than $30 billion, but also from the near term prospects of self-driven vehicles from the likes of nontraditional competitors like Apple and Google.

In financial services, traditional retail banking customers are opting for the convenience and choice of online services while the number of physical retail branches are contracting. The payments industry is seeing the effects of mobile payments. Over the longer term, digital currencies will enable peer-to-peer payments and micro-finance services, which greatly simplify complex financial arrangements.

Great potential exists to address even the most complex challenges like wearable technologies to transform healthcare, online education and self-paced learning to rethink education, and the integration of sensors with machine learning to address the complex challenges of the environment. To our most innovative thinkers, the possibilities seem limitless.

While the application and pace of change may vary widely by industry and region, some core principles appear constant:

1. The barriers to entry for both local and global markets have reduced dramatically as the rapid adoption of the cloud, particularly infrastructure as a service (IaaS), has made access to IT capacity inexpensive and massively scalable.

2. The global adoption of mobile broadband, combined with social media and ubiquitous messaging, has redefined customer, reach, engagement, distribution and customer service.

3. The integration of connected endpoints with highly scalable compute resources offers the promise to streamline a multitude of industrial and consumer processes.

4. Capital, particularly in developing markets, is readily available as investors reward innovation and growth.

5. Globalization requires companies to both innovate with new products to achieve and become low-cost operators to sustain market position.

The key to unlocking the value of these technologies remains the ability to not only collect and interpret intelligent streams of data, but also to assemble components to quickly create new sets of capabilities. For example, smartwatches will gather larger streams of personal data, which health care providers will utilize to provide an increasing portfolio of services.

It's all about integration. Not just the integration of hardware, applications and data, but the integration of ideas, internal teams, external partners, and new business processes as well.

There is an emerging set of requirements to enable different organizations to connect applications and data freely across traditional boundaries. This ability to expose and make data open for reuse wherever and whenever it's needed creates the "adaptive enterprise" of the future. For example:

- The software-defined data center is the coming wave where lower-level physical components are used to build powerful, highly-scalable systems at dramatically reduced costs. Amazon used a new design model for computing to move from its merchandising roots to building a market-leading service for delivering infrastructure as a service. It now exceeds an $8 billion annual revenue run rate.

- Cisco estimates there will be 50 billion devices connected to the Internet by 2020 and is investing to bring secure connections in fields ranging from energy measurement to connected cities.

- GE Chief Executive Jeff Immelt has set a target of $15 billion in software revenue by 2020, partially enabled by a new operating system that will create a common platform to effectively ingest and process machine-to-machine data.

- IHS Automotive says that 20 percent of vehicles sold worldwide in 2015 will have included some form of embedded connectivity, with the number growing to 152 million by 2020.

- Google bought smart thermostat maker Nest Labs last year for $3.2 billion, and Samsung purchased connected home company SmartThings for $200 million.

As technology leadership once again transitions to enabling business competitiveness, IT leaders must adapt from their traditional roles of providing and securing infrastructure to becoming an active champion of innovation. At the same time, they must execute capabilities that ensure new forms of information can be integrated with advanced analytical capabilities. The role shifts from Chief Information and Chief Security Officer to Chief Innovation and Chief Integration Officer.

In my thirty-plus years of experience in the technology industry, ranging from development to CIO to CEO, I have participated first hand in the major transitions that have moved markets. At no time have I been as excited and optimistic as I am now about the role of technology in creating value and improving people's lives.

Several years ago, I met the leadership team of MuleSoft and I now serve on their Board of Directors. I have watched first hand as the company has built the market for developing and deploying open APIs. They've created a new way to integrate these new technologies and to help companies in a range of industries bring innovative solutions to scale.

The reality of a connected world is here and now. Technological innovation is a core business strategy and the roles of a company's strategy and technology leadership have become explicitly interwoven. Tomorrow's leading companies will live by the creed of "disrupt or be disrupted."

Michael Capellas was previously Chief Executive Officer and Chairman of VCE Company, LLC, a joint venture between EMC and Cisco with investments from VMware and Intel. He was Chairman and Chief Executive Officer of First Data Corporation and was Chief Executive Officer of MCI, Inc. through its 2004 restructuring and acquisition by Verizon in 2006. In addition, Capellas served as Chief Executive Officer and Chairman of the Board of Compaq from 2000 until its merger with HP in 2002. Post merger, he served as President of HP.

Part I

The case for change and the opportunity for digital transformation

CHAPTER 1

Introduction: Enterprise Darwinism

by Ross Mason, MuleSoft

Business as we know it is under enormous pressure, and nowhere is that pressure more pronounced than in the IT organization. The customer experience has completely changed. The expectation of what they should get from their vendors has changed as well. Everything needs to be dynamic, always on, and always mobile. Customers expect personalized service. And if they don't get what they want, they will move on and take their business with them.

But that isn't the only pressure point on the modern business. The barrier of entry into almost every market has evaporated. With major disruptions happening in financial services, retail and healthcare, no company is too big to have the rug pulled from under it. It is no longer about the big eating the small; it is now about the fast eating the slow.

In response to these pressures, businesses are turning to technology to help them move faster and deliver more to their customers. There is an increased need to bring in best-of-breed cloud applications; for mobile applications that drive better customer experiences and better, faster ways of connecting with partners and suppliers; and for agile and adaptable processes that help employees be more productive. Every function in your organization is lured by purpose-built cloud applications that help them do their jobs more efficiently. The role of IT and integration today is, "How do I get these connected quickly without making a mess? How do I move information between these different systems to better run my business?"

The reality for every established enterprise is that the centralized IT model is no longer working. Technology is no longer a central concern. Software is used in every corner of the business to perform all kinds of functions. This is putting tremendous stress on IT as the business makes more and more requests to get its new application to work with other applications and data sources. This creates a massive IT delivery gap.

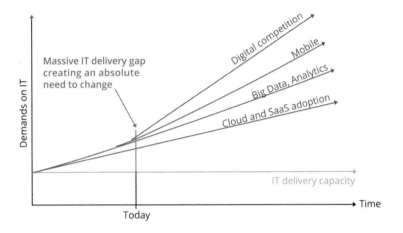

IT has to figure out how to deliver much more to the business without investing a lot more—and always ensuring a positive ROI on investment. IT has to devise ways to make its cost of operation, both in terms of money and time, lean and efficient. IT has to deliver the capability to support these new channels to the business, and to do so must become expert at extracting value from all its assets locked up in dinosaur systems: those mainframes, those legacy applications, those databases, and big data stores. In and of itself, that's all very hard to connect, but it's been made even more difficult to connect with the advent of the cloud. IT cannot afford to get bogged down in technical debt trying to deliver everything to every area of the business. Instead it needs to enable and empower the business to autonomously deliver their own projects while retaining governance and control over mission-critical data. Modern connectivity isn't about creating integrations to piece together new projects; it's actually about how you take that value—those assets buried inside your data center—and surface it to the front of the business to enable the application developers to create apps, reports, dashboards and business process. How do you connect those assets to your audiences: your customers, your partners and your employees?

If that wasn't enough, those startups that are, or about to, eat away at your core value propositions and your growth plans aren't dealing with a legacy. They have become very good at pulling together cloud applications, public APIs, and open source components to build companies and products at an alarming rate. This composable mode of operation is normal for startups. It's very abnormal right now for most enterprises. That's the competitive landscape that you're dealing with. And this kind of digital Darwinism is extraordinarily unkind to enterprises who aren't prepared to deal with the realities of the new world they live in; increasingly we are seeing winner-take-all markets consolidating revenue at the top of every category. As Charles Darwin famously said, "It is not the strongest of the species that survives, it is the one that is most adaptable to change." Every industry in the world is affected by the digital revolution, and every company is looking to figure out how they can cope with the change and take advantage of the opportunities it affords.

Getting great at change

One of the big changes that every CIO has to accept is that IT is no longer just about keeping the lights on and keeping the network running. The role of IT is not about delivering projects for the business; it needs to be an enabler to the business, and must enable them to create value for its audiences. That means IT needs to deliver platforms and capabilities to allow the business to self-serve data and resources. A business analyst should be able to find real-time data sets in a data marketplace and an application developer should be able to provision a new application environment in seconds. That's a different way of thinking about what many CIOs do today. How can an IT executive actually shift that mindset from project to project delivery inside your organization to opening up the assets and capabilities of your organization? You can do that with an approach we call *API-led connectivity*.

In the modern world, everything is digital and everything needs to connect. API-led connectivity is an approach to organizing your business from a technology perspective. Connectivity is no longer about just integrating back office systems. You need to find ways to unlock the value of your data and assets to allow the business to actually take part in creating the new digital experiences that it so desperately wants to drive to their consumers. Your technology assets exist everywhere—in files, databases, applications, warehouses, big data stores in the cloud and on premises. Cloud applications are just another enterprise asset; they're another source of information that must be built into your IT landscape. The way you connect to those customers, partners, and employees is through applications. By applications, I mean web applications and mobile applications; even the Internet of Things (IoT) is driving new experiences for consumers with connected devices. The IT organization needs to focus on providing the building blocks for the business to build the applications, reports, analytics and digital products to compete in this digital era.

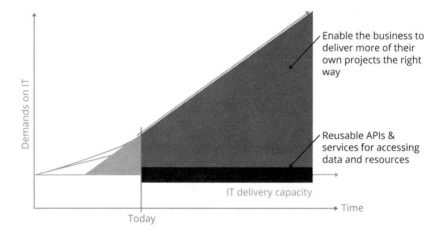

The business is going to build these applications with or without IT. The role of IT is to give the business consumers capabilities - reusable APIs - to get access to enterprise assets the right way. The problem with the business going around IT is that it creates major downstream problems that strangle agility because systems become tightly bound together and cannot be changed without months of work.

Jeff Bezos, the CEO of Amazon, understood this problem at his company and created a mandate to stop different teams doing the wrong things. The now famous memo, Jeff Bezos' Big Mandate,[1] said:

---

[1] Stevey's Google Platforms Rant first published Jeff Bezos' Big Mandate:
https://plus.google.com/+RipRowan/posts/eVeouesvaVX

1. All teams will henceforth expose their data and functionality through service interfaces.

2. Teams must communicate with each other through these interfaces.

3. There will be no other form of interprocess communication allowed: no direct linking, no direct reads of another team's data store, no shared-memory model, no back-doors whatsoever. The only communication allowed is via service interface calls over the network.

4. It doesn't matter what technology they use. HTTP, Corba, Pubsub, custom protocols — doesn't matter. Bezos doesn't care.

5. All service interfaces, without exception, must be designed from the ground up to be externalizable. That is to say, the team must plan and design to be able to expose the interface to developers in the outside world. No exceptions.

6. Anyone who doesn't do this will be fired.

Looking past the aggressive governance outlined in bullet 6, one of the most important statements here is the third bullet: "no back-doors whatsoever." Those back-doors are what disable agility and innovation. Back-doors become trapdoors for any new initiative because they create invisible dependencies between apps, and often perform unnatural acts to get the data or resources. Backdoors are created if the business cannot get what they need through the IT channel. IT needs to change the way it operates to unlock the data and resources need by the business and ultimately allow the business to self serve while IT governs access to the data.

IT can't own everything

Another major challenge is the sheer number of enterprise applications due to the explosion of SaaS applications, mobile applications, and the coming of IoT. The number of things the business can use to improve the way they work means IT can no longer own all the applications. This is dubbed "shadow IT"—applications procured outside of central IT—and it is a natural progression as every corner of your business needs these applications to work more efficiently. However, IT needs to own and govern the data that powers all those applications because IT is still on the hook for data security and governance. If there is a data breach at an organization whom does the CEO hold accountable? Yes, IT is still accountable for all data even if the application that was compromised was built or procured without IT.

Rather than owning every piece of the application stack, how does IT unlock the value of that data so that other people in the organization can build those applications without compromising security? If you look at who can actually build apps in the enterprise today, you have designers, business developers, mobile app developers, and development partners; they're all going to build new applications and new experience for consumers with or without IT involvement. What IT needs to do is think about more about governing the data and providing self-serve access to that data.

To do that, you need to be what is called "composable." That means people need to be allowed to self-serve, grab pieces of information and bits of functionality that they need, and allow them too build these new experiences. Your IT today is like a reliable bicycle. It works, it gets you from A to B, and it's not broken. But the rest of the business wants to create different experiences for their consumers, and they want to experiment here. They want to evolve how they use the assets inside the organization. Delivering one thing to them no longer works.

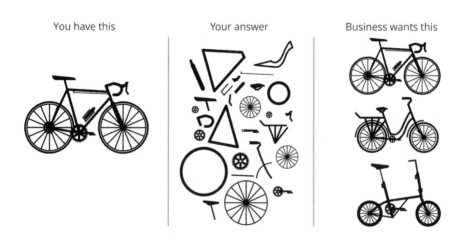

You have this   Your answer   Business wants this

What you need is this. You need to have the component pieces of your environment as composable APIs — composable services — that you can pull together in different ways to experiment, evolve, and drive a different type of outcome for you and your customers. The way we propose doing this is with API-led connectivity.

API-led connectivity: A straightforward approach to delivering a new operating model

API-led connectivity is a straightforward approach to delivering this new operating model. The API-led connectivity approach provides a view of your organization based on your reusable capabilities and who should own and can access those capabilities. It's a different way of thinking about the problem. The ownership and accessibility shifts because you are actually trying to unlock the value of your organization by decentralizing access and value creation throughout the business. The API-led connectivity approach takes a three-layered model with a System layer, Process layer and Experience layer. These layers open up data, model processes, orchestrate and augment data, and help you define innovative customer experiences. The following chapters in the book will go into detail about each layer and how they work together. This is a very powerful approach because it drives a fundamentally different type of application development experience inside your organization.

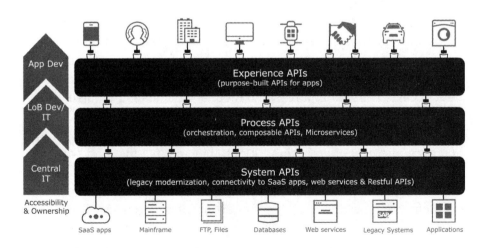

The role of each layer is to provide access and governance to the assets below. Each layer can deliver a different consumer experience so that you can enable people with different skill sets to access and leverage capabilities. Imagine you have a JavaScript developer, 22, fresh out of college, joining your company. She knows how to build great consumer applications. If she has to understand all the complexity of the back-end systems inside your organization, she is never going to build that rich consumer experience. In fact, you're going to have a hard time holding that talent in your organization. That's why API-led connectivity isn't just an architecture, it's a way of operating, a cultural shift—a shift in the way IT behaves and in the way the business behaves. IT delivers capabilities to the business that support a broader range of skillsets; in turn the business learns to self-serve and not work around IT. The value of these layers is to relieve tension, retain control and to allow other areas of the organization to be successful, autonomous, and drive new value creation both inside and outside of central IT. That's really what the API economy is about. It takes what already exists and building new value on top of it. You want to build and enable the API economy mindset inside your organization.

Why the API economy matters

The API economy is simply the sum of digital exchanges between two or more parties. The Web has rapidly shifted from being about exchanging information to exchanging value and powering a whole lot of new thinking around scalable business models. Disruption is happening across industries as startups figure out how to connect audiences with the things they need and want in a faster, consumer-centric way. This is why Airbnb can rise as the world's largest supplier of guest rooms without owning any real estate or why Uber doesn't own a global fleet of cars. These companies and thousands of ideas that died before liftoff are the product of mass experimentation on the Web. Startups can now go from zero to hero in a matter of months if their timing, audience and execution is right; if not, they just fail fast and someone else has a go. Open APIs on the web are a key driver here because they make everything from analytics to prediction engines to historical government data available at the fingertips of millions of developers worldwide. Every enterprise needs to replicate some of that capability within their organization; they must unlock the value of their data and assets so that the whole organization —and sometimes partners and external developers — can experiment quickly, effect change quickly, and fail quickly.

If you are even just half-successful at doing this, your business will reap significant rewards. Some of our customers are using the API-led connectivity approach to drive their mobile initiatives. Some are using it for unlocking their assets internally to break down silos between business units to drive more value and more innovation from the inside out. Increasingly we're seeing more enablement for IoT as well. IoT, APIs, and connectivity go hand in hand. While APIs for IoT are often considered the way devices end up passing data back to operational systems, putting productized APIs on top allows other developers to leverage the data that's being generated. It means that other people can come in, discover the data, and access the data the same way for their applications. In short, APIs mean leverage. You're actually unlocking the assets and exposing it to new audiences so they can build value on top of your value. That's how the API economy works. That is how the Web works; that is how the ideas behind Airbnb, Amazon and Uber are born.

If the business is organized around the API-led connectivity approach, connectivity tissue grows to provide a single point of management, governance and visibility, and IT gets an unparalleled view of your enterprise. What you have now is not a big collection of applications and connections between them, you now have an *application network*. Your important information is available on the application network as discoverable, self-describing APIs. New data and resources can be made available quickly because you now have a consistent process to build and publish APIs on top of your applications and data sources. You can now see exactly what's going on at every point, from the point when a request comes in through a mobile app all the way down into that mainframe system that nobody wants to touch. That's unparalleled visibility.

The "business as we know it" wake-up call

The way you win in this digital era is by having good services and how well you can integrate the experience between different functions. You win on how well can you hide the complexity of your legacy to deliver a delightful consumer experience. Uber has been the wake-up call here, but it's not the only example of this type of disruptive business model built with rapidly composable technology which capitalizes on this new approach to technology, people, and processes. Companies like Airbnb, Alibaba, Apple, Google, Salesforce, FitBit, Stripe, Netflix and Amazon, which have become platforms of digital capabilities, have shown us that the ability to compose your assets into new products and services allows you to quickly change in reaction to the market needs. Your business becomes a platform when you can recompose easily through APIs digital products and services.

In 2012, Uber was a great taxi service. In the last few years, though, they've literally exploded. They've expanded faster than many other companies in history. In the beginning of 2015, Uber opened up an API to third parties, and immediately it switched on a whole new set of digital channels for them. Now, if I go to the United Airlines app, when I go for my boarding pass, I can also book an Uber to the airport. If I go to Google Maps and I do a search, one of the modes of transport is Uber. That's unbelievable. Google gets billions of hits a day on Maps, and now Uber is plugged into that value chain. In a matter of months, other companies have started sucking in the value that Uber provides into their application to create new value. This is the way the enterprise needs to operate as well. You can't operate in silos. IT cannot be centralized. It's got to be decentralized to enable more of the organization to do more with the core assets and deliver more value to your customers.

Where do you start?

The challenge here, of course, is how you get started. API-led connectivity is all about opening up your data and allowing the people who can use it to self-serve (those people will often work outside of central IT). All this happens while IT retains control and governance of that data so that you have security, data quality and governance. That's critically important. The API-led connectivity approach defines an operating model that helps you think about how to organize your assets inside your organization, and who should own those assets, and what are the roles of those assets. The API-led approach defines System, Process, and Experience layers. They serve different audiences with different skillsets performing different functions. This approach to organizing reusable assets helps you to align to the needs of your different consumers in a purposeful way. It also means the assets can be delivered by different teams running at different speeds. While going faster is the goal here, you don't want System layer APIs changing as frequently as APIs in the Process layer.

This book is designed to help you think about what API-led connectivity would mean inside your organization. How will you start to change and redefine the role of IT in your organization? How do you deliver those reusable building blocks to the business? How will you start to become a composable enterprise? What are the measurable outcomes? How can you create an environment open to change? Change isn't easy, but it is necessary. This is enterprise Darwinism. If you start down this journey, you will find that your customers are happier, you can open up new streams of revenue, and grow faster overall.

I look forward to taking this journey with you.

CHAPTER 2

The application network:
What it is and how it works

By Ross Mason and Uri Sarid, MuleSoft

We are in the "fourth industrial revolution"[2] where the forces of digital disruption deeply affect businesses of every size and in every industry. It's an era where all parts of the enterprise function using software, SaaS, and custom apps. More than ever, consumers have lofty expectations and are willing to shop around. And if that wasn't enough, the barriers to entry for each industry have evaporated. Startups everywhere are disrupting all types of products from financial services to crowd-sourced insurance to customizable health care to unique shopping experiences. A new idea can be built in a hackathon and developed in the garage using major building blocks like unlimited cloud compute power, free dev tools, open APIs, and open source components. These new entrants build fast, move fast, and continue to educate consumers on the type of experience they should be getting. Consumers, employees, and partners now expect everything to be on demand and fully connected.

There is not a single business in the world unaffected by these changes, but large organizations with calcified processes and monolithic infrastructures are hit the hardest. Fortunately, the means for real business transformation are within reach. An application network can help an organization be more agile, change the clock speed of business, and respond to changes in the market, the business, and the consumer without having to break things.

---

[2] Wladawsky-Berger, Irving "Preparing for the Fourth Industrial Revolution." *Wall Street Journal*, Feb. 26, 2016.

Companies need to launch new solutions that innovate in order to provide new products and experiences to meet customer demand, get ahead of the competition, and make their businesses more efficient. Actually producing these solutions, particularly in larger, older organizations with a technological or cultural legacy, can be difficult. Customers, employees, and partners are demanding ever faster innovation; if a business can't deliver, there is a real risk of losing customers, surrendering market share, and missing growth targets. MuleSoft conducted a recent survey of 800 IT professionals and found that 66 percent say they are under "drastic" pressure to deliver technology projects across the business; a similar number say that change has to occur[3] in order to meet business demand.

The problems are compounded further by the increasing number of people operationalizing change, particularly those not in IT. Businesses who are using and deploying technology in traditional ways cannot respond quickly to change. In addition, the creation of and alterations to monolithic business processes is time consuming and complex, creating paralysis rather than agility.

IT cannot keep up with changing business needs if it continues on a traditional path. If IT continues to work as a sole technology provider for the business, it can neither integrate systems quickly nor leverage the big data now coming from all parts of the company's operations. Radical transformation can happen if you create smaller, agile units for business, processes *and* underlying technology, while still maintaining stability and control.

To effect the kind of change that will allow organizations to compete in a truly disruptive business environment, we propose a new vision of enterprise IT: the application network.

What is an application network?

Application networks seamlessly connect applications, data and devices. They take a different approach to the methods used to connect applications, data, and devices today. Instead of utilizing point-to-point connections or isolated architectures, the application network provides an

---

[3] MuleSoft Survey Finds Top Five Reasons APIs Are Taking Centre Stage for Business. January 12, 2016.

infrastructure for information exchange by allowing applications to be "plugged" into the network. The network can be as simple as two nodes that enable two applications to share information, or it could span across the enterprise as well as to external ecosystems. Designed to honor Metcalfe's law, every new node added to the network will increase the network's value since the data and capabilities of that node are made discoverable and consumable by others on the network.

Unlike prior approaches to connecting applications in the enterprise, an application network is designed to allow many people inside and outside the enterprise to have controlled access to valuable business data. This will occur by allowing anyone in the business to use consumption models they are familiar with. In other words, it makes it easier for someone in the organization to create a useful application, use of data, or an API creating a particular experience, and then expose that value to the network.

For example, someone might have found a way to expose the organizational structure of a group of people in the company, perhaps by tapping into the HR services provider system the company uses to manage its employees. To do this, they may have used the HR services provider's API and some credentials to pull in the org chart below a particular executive or by function. Or they may have wrapped that HR services provider's native API with a friendlier API optimized for consumption in certain use cases. They likely have done so as a part of a bigger project: perhaps to create a portal of skillsets, or to manage access to source code repositories, or to synchronize with a disaster recovery contacts database that needs to be available offline.

If this work is made available beyond the scope of this project — e.g. by making the friendlier API available to others, or allowing the connection credentials to be reused, or by capturing the project in a template, then the provider's service is exposed to the network and may be leveraged in other scenarios. Some of those use cases might be a team needing to create dynamic visualizations of resources applied to company objectives, whereas another team might need to automate the laptop requisitioning request process. Through the application network, they can more easily access the HR services provider's employee subsystem through the work of the original team, without the explicit help of that team.

The key aspect of an application network: reuse

The key aspect of an application network, is reuse; a node of the network must be reusable elsewhere via its connection to the network. For instance, in a social network, the information of a member — whether in the form of the content they post, or in their rating of existing content (explicit or implicit), or in various other signals they provide — is then reused across the network. A social network member's content is viewed by many others; that content can then be shared with others on the network. Those ratings and other signals inform what content is promoted to yet more members; in other words, they are reused by the larger subset of the network. The network enables reuse by other members whether those members understand their work has been reused or not. This induced cooperative behavior provides the classic network effect, and is the core source of value of the network itself, atop the value of each application to its direct consumers.

As in other networks, the value of exposing information from and allowing access to the application network comes at a price; not all information should be exposed to everyone, and not all capabilities should be made available to just anyone. In a social network, members may restrict certain content to be viewed by only some of their connections; they may only allow certain people to follow them, or to post to their timeline. Attention must be paid to the transitive nature of this access. If I grant access to a friend, and they grant access to another friend, does that friend-of-a-friend (FOAF) also have access, perhaps on a more limited basis?

For an application network in the enterprise, this manifests via APIs and the transitivity of API access. For example, an employee management system API allows access to employees and their data based on entitlements assigned to the API's clients, and perhaps to the end user on whose behalf the clients are accessing that API. The team that integrates with the system API will need to register as a client of that API, and they will need to figure out what entitlements to set up and establish which end user(s) on whose behalf it will access that system API. And if this team in question offers its own API which makes use of the system API, it will need to think carefully what access clients of its API should have to the employee management system API.

Success depends on self-interest

Another key aspect of the application network, and one that affects the viral growth and stickiness of the network, is the degree to which it caters to the self-interests of each node. Facebook, for example, depends entirely on its ability to satisfy the innate curiosity of humans about other humans; in particular, it satiates their need to know what their friends (real or otherwise) are doing and offers engaging content to each member based on their taste as inferred from their friends' tastes. Any given member is engaging with the network primarily because of their self-interest, not because of an interest in contributing to the network. Members curate content for their friends knowing their friends will see it; they will judge their friends' reactions and rate their friends' content, which establishes reputation and self-image. It is then up to the network and its dynamics to translate that to the value of the network as a whole.

This is equally true for an application network, in which the teams attaching the applications and creating valuable new services are usually most concerned about getting specific projects done, rather than creating reusable assets for the whole organization. A successful application network must find ways to align the teams' self-interest with the creation of reusable assets that benefit everyone. Just as a social network must allow members to keep some content private and some access limited, an application network must also allow some services, design assets, and APIs to be kept private, without necessarily imposing on the teams the burden of maintaining them for a broad population of other teams to reuse.

Many factors can influence this balance between the autonomy and independence of each team and the value of their work to the broader community. There may be incentives placed to encourage building reusable assets, such as recognition by the community or management; constructive criticism that goes to increasing the quality of the asset; access to other services ("you can use my service if I can use yours"); funding of some sort to offset the cost of maintaining reusable assets, etc. In some companies, the reusability of assets may be imposed as a requirement. For example, a review process may be instituted to ensure all integrations and APIs that access an HR services provider are documented and registered as potentially reusable assets, and are reviewed for compliance with security and privacy concerns, usage limits, and so forth.

Whether assets are in fact reused will also depend on the self-interest of teams engaging with the network and consuming or contributing to it. In traditional enterprise IT initiatives based on Service-Oriented Architecture (SOA) principles, reuse was mandated from the top. Certain patterns and service interfaces and architectures were predetermined by a small group of enterprise architects and were required to be reused by all teams. While this approach may still make sense in some scenarios, increasingly it is being replaced by "merit-based reuse": potentially reusable assets are created by a broad population of teams in the course of implementing projects; those assets are then consumed or not by that same population based on the merit each team finds in them. This approach prefers the "wisdom of the crowd" over that of the central architects, who often participate by guiding or enabling the teams and potentially pruning assets rather than pre-determining everything.

An analogy may be found in the world of mobile consumer apps. Rather than the traditional app stores based on a single telecom provider's creation and curation of mobile apps, the modern app stores allow apps from a broad base of developers and let consumers "vote with their feet" to determine which apps have more merit. Usually, as long as there is a vibrant community of both consumers and producers, the best apps rise to the top rather quickly and are reinforced in a non-linear way. The result is a much healthier, more valuable, and self-sustaining ecosystem of apps than would have ever been produced by a single controlling provider. The role of the provider, and of the application network architects, is to provide just enough enablement to seed the network and to enable and encourage best practices and just enough governance to create a trusted ecosystem and safety net for the organization.

Benefits of an application network

In order to serve the needs of the business and provide the benefits that organizations need to thrive in today's connected, competitive environment, an application network must be:

- **Clearly defined:** A well-defined operating model enables the business to get the most value from the IT assets on the network.
- **Broadly deployed:** It empowers the whole organization — not just IT — to discover and access the data and assets to fuel applications, reporting, analytics, business processes and automation.
- **Accessible and modular:** Every application is accessible and recomposable to fit the rapidly changing needs of the business.
- **Seamless:** It is a seamless network of applications, data sources, devices and APIs, whether in the cloud or on-premises, home-grown or acquired.
- **Elastic:** The application network is elastic and can grow or shrink depending on the demands of its consumers.
- **Scalable:** It allows companies to introduce new applications more easily because the application network already defines how to access, expose and share new application data.
- **Secure:** It can be organized to reflect the security and governance constraints in a business of any size.
- **Monitored end to end:** Data running through the network can be tracked end-to-end, monitored and analyzed. The impact of a request from one node in the network can be traced back to all the other nodes that take part in that request.
- **Analytical:** The network understands dependencies between applications and can perform impact analyzes on changes made to applications on the network. This gives organizations a real-time enterprise wiring diagram that can be inspected.
- **Measurable:** The value of an application network can be explicitly measured through a set of built-in KPIs that track reuse, utilization, adoption and redundancy.
- **Organic growth and reuse:** As the first applications are connected to the application, data and functionality is made available for reuse. IT and the business can plug in more applications over time. Each application adds more value to the network, more reusable building blocks for building applications, services, and business processes. The business and IT can use the existing assets to build new assets which are also made available through the application network.

Capabilities of an application network

When an application network is in place, it allows businesses to break down information silos as well as build and deploy apps and processes faster, all in a highly scalable manner. It creates room to govern and control while allowing the business to reconfigure the architecture easily.

The qualities of an application network allow for these capabilities:

- **Standard communication:** The base architecture for the application network provides a standard communication model which simplifies data exchange between applications on the network.
- **Security:** Any application connected to the application network will be subject to the same security policies and access controls. Different domains can have different security policies, which allows enterprises to segment and provide access to that data depending on the data source, consumers, geographic location, or other factors.
- **Reusable, modular components:** All data resources on the application network are made available through reusable interfaces, meaning that any application on the network is composable.
- **Layered governance:** The application network mandates well-defined interfaces to access resources. It also provides a layered governance model that makes it possible to track data consumption from consumer applications (such as mobile apps, web apps, dashboards, analytics) all the way through to the back-end data. It can also track dependencies between applications and even perform change impact analysis.
- **Discoverability:** The application network is designed to exchange information between applications and people. This means that assets on the network are discoverable. In addition, different consumption models can be designed around different groups of users within or outside the enterprise.
- **Custom consumption models:** It provides a set of core APIs, allowing solutions to be built that enable different consumption models for different groups of people across the enterprise

(such as a "data marketplace"). This might be useful for business analysts to power their BI reporting.

- **Self-service enabled:** It provides a core set of services to enable consumers (developers, analysts, data scientists, creative teams, mobile developers, ops and admins) to access the network in ways that makes sense for them through tools they understand. It also has a set of APIs that allow new consumer models to be built.

The key design principles of an application network

When considering the genesis of an application network, it's important to define the design principles that shape how its components are built:

- **The network is opinionated:** The network has built-in, well-defined and systematic mechanisms for attaching new nodes and enabling them within the network. These include publishing APIs and their specifications, registering client access grants and hence also client dependencies, bringing them within a security context, etc. It also provides built-in, opinionated ways of operating the nodes and tending to their life cycle, of collaboration between users on the assets and their lifecycle, and of analyzing and securing the network. The network is not intrinsically opinionated about the specific architecture of which applications must be connected in what ways, as that is often determined in an incremental, merit-based way, vs a top-down traditional SOA approach. However, architects may suggest or impose certain architectures, API patterns, standardized data schemas, etc. to enable and potentially constrain how the network is built.
- **The network is consistent:** Rules and policies are enforced within a domain. Artifacts such as APIs, templates and portals inherit the services and restrictions of the network. This means that APIs are subject to the policy enforcement of the network; templates use common error handling defined by the network.
- **Applications are connected to the network:** Conversely, applications are mostly not connected directly with each other. The connection into the network contains rich metadata about each

application which the network uses to expose that application inside the network.

- **The network adheres to Metcalfe's law:** Adding a new node to the network must add n+1 value to the network. That value must be a set of KPIs that are built in and can be tracked.
- **All data connected to the network is discoverable and addressable:** Every resource is exposed through through system APIs. Not all data will be exposed to consumers — this is an administration choice.
- **The consumer is king on the network:** The fundamental value of the network is to unlock and enable consumption of data and resources. Every person in an enterprise can connect and get value from the network. This means the network must cater to different consumer profiles across technical and non-technical roles.

How is an application network different from SOA?

The principles of SOA still are valid today, but the needs of the organization have changed drastically, requiring an equally drastic shift in how IT works with the rest of the business. SOA had to be built from the ground up for every company; it therefore became a reflection of the complexity and custom applications that defined the enterprise's IT landscape. In contrast, the application network is designed to enable the whole business, not just IT. This is key, as innovation cannot occur at the necessary speed if IT is the sole technology provider to the business. The needs of the business will grow faster than the capability of IT to meet them. Finally, SOA is an architectural concept, while an application network is a tangible product that connects not just applications, but people and processes as well.

It's important to note that services must be loosely coupled to provide the flexibility and scalability that businesses need. Tightly coupled networks become brittle and expensive to maintain, which defeats the self-service ethos of the application network. Simply completing a lot of integration projects in a traditional way does not result in an application network nor does it provide an application network's benefits. Instead of connecting applications through point to point connections or isolated architectures, application networks provide an infrastructure that can achieve reuse of services, knowledge, access and best practices.

This means that IT has to shift its mindset from complete control to a distributed authority over technology in the business. It requires a federated approach to governing and controlling data, applications, and systems. This allows IT to be a platform enabler for the organization, will increase productivity through reuse, and will make change more predictable and easier to manage.

Conclusion

With the massive number of applications, data, and devices that need connecting in the modern enterprise, and the incredible amount of time and resources that companies spend trying to tie everything together, an application network can provide the agility, flexibility and speed that businesses in today's environment urgently need. New applications can be plugged into the application network as easily as you can plug in a printer. The application network can deliver unified vision and control and offer intelligent data about the relationships between different applications. A new vision of what your IT organization can be and do can allow your business to harness the digital revolution; an application network is designed to make that happen.

But achieving such a vision takes a new kind of leader—a leader who sees herself or himself as a business partner rather than a denizen of the back office; someone who understands the strategic importance of integration, connectivity, and the need to radically shift how businesses understand and deploy technology. To realize an application network, you need a new kind of CIO, and we'll be meeting that person in the next chapter.

CHAPTER 3

The CIO as Chief Innovation Officer

by Joe McKendrick, Forbes staff writer

"Right now, your company has 21st century Internet-enabled business processes, mid-20th century management processes, built atop 19th century management principles."

—Barry O'Reilly, principal at Thoughtworks and co-author of *Lean Enterprise: How High Performance Organizations Innovate at Scale*

The organization as we have known it is barely clinging to life support. Increasingly, the calcified, creaky business systems and processes that have been embedded in organizations for decades are being swept away by more agile digital disruptors. Online and app-based companies—powered by cloud, open APIs, data analytics, mobile, social, and connected to the Internet of Things—are redefining markets and raising consumer expectations.

A recent survey published by MuleSoft confirms this trend.[4] Established businesses and startups alike are responding to this new digital environment by structuring themselves as *Composable Enterprises,* built out of connected software services, applications and devices, the survey reveals. Composable enterprises are connected organizations with business processes supported by on-demand services that are acquired and leveraged from the cloud and APIs, furnished by outside providers or through internal data centers. These services are connected through APIs.

---

[4] MuleSoft Connectivity Benchmark Report. July 2015.

Composable enterprises are able to create digital competencies that can be leveraged by others, thus extending the bounds of their enterprises. For example, a company such as Amazon has extended its core competency to being not only an online retailer, but also an IT solutions provider through Amazon Web Services.

To achieve composability, most companies are digitizing every aspect of their operations. "Digital disruption is more than just a technology shift," says Ray Wang, founder and principal analyst at Constellation Research. "It's about transforming business models and how we engage." And, as Wang points out, "digital disruption is creating winner-takes-all markets—the past five years have shown the difference between those who invested in digital transformation and those who have not. The corporate digital chasm is massive among market leaders, fast followers and everyone else. Look at Amazon versus Borders, or Netflix versus Blockbuster. Digital Darwinism is unkind to those who wait."

Taking charge of this digital transformation requires an in-depth knowledge and understanding of the new systems and processes that are driving new businesses, how they interact and what they are capable of delivering. This transformation will deliver more than efficiency and lower-cost processing—it will bring heightened competitiveness through digital engagement and analytic decision making.

Chief Information Officers and other IT leaders, with their extensive understanding of the power of technology to deliver business results, are a natural fit for this role. Many CIOs recognize the challenges and opportunities ahead in bringing new technologies and resources into their organizations. "Every part of the business needs technology," says Ross Mason, founder and vice president of product strategy at MuleSoft. "Every conversation about anything in the organization has technology featured in some way." This is creating new and challenging roles for IT leaders; as CIOs increasingly take on the role of chief innovation officer, they are expanding their roles into new realms.

Joseph Brophy, solutions development manager for New Zealand Post Digital, knows firsthand how technology can move the business forward. This growing service, offered through New Zealand's postal service, is led by technology visionaries to elevate the digital customer experience and accelerate growth in ecommerce and logistics. "We're tasked with searching for new business models and bringing new products to market," Brophy says. To accomplish this, New Zealand Post Digital has built and makes available APIs that provide connectivity to the organization's back-end processes. And it all begins with IT. Brophy notes, "IT is a key enabler of our digital customer experience. It allows us to stay close to the market."

The composable enterprise, led by the Chief Innovation Officer, moves quickly to adapt to changing customer realities and requirements, and is not dependent on any single type of technology. In the past, the IT leader's role was reactive, ordering and building systems to meet requirements, preparing and delivering reports, and troubleshooting systems issues. The new IT leader's role is proactive, encouraging innovation and even acting as an agent of disruption for his or her business. Accordingly, we have seen a dramatic shift in the role of IT in the business. CIOs are emerging from the confines of back offices and server rooms to serve as high-level consultants and even digital provocateurs, shepherding their businesses through the disruptions that are now turning entire industries upside down.

Tom Quinn, the CIO of News Corporation Australia, is one of those digital provocateurs. "The disruption in the media industry was easy to see," he says. "But it took some time for companies like News Corp to see that this huge shift was coming. It took a number of years of losing business for the light bulb to go on to figure out, okay, this business is in a bit of trouble. The problems I'm looking to solve today are many, but all related to a business in crisis. If we don't transform, if we don't get better at delivering our content across multiple platforms, then we'll go out of business." Quinn's experience suggests that along with providing traditional technical capabilities, IT also needs to take a leadership role with the business, guiding their organizations through the fast-growing digital economy. "It's no longer just about keeping the lights on, making sure the network runs, making sure one has the right software on the desktop," says Mason. "CIOs have a whole new set of challenges, such as supporting marketing to build new mobile applications that engage consumers in new ways, or supporting HR to build applications that help them work more efficiently." This means that

CIOs must emerge as a partner to the business, rather than remain in a support role. Here are the key attributes of emerging composable enterprises, and the new roles CIOs—as chief innovation officers—will play in helping to lead them:

- In the composable enterprise, IT is an enabler. All too often in organizations, IT has been the department of "no," or the place where project requests go to languish. In a composable enterprise, IT serves as a full partner to the business, but at the same time, may even be invisible. Services rendered through the cloud, APIs and mobile are part of a structure that enables decision makers to quickly access the applications and information they need. "It's not just about connecting applications together, it's also about exposing information through things like APIs that enable people to be able to grab information easily, in a self-service way, without having to go through central IT for everything," says Mason. "This bolsters the reputation of IT of becoming a partner to the business."

- Leading a composable enterprise is a team effort. Technology leaders need to step into business leadership roles, just as business leaders are embracing technology roles. CEOs, CFOs and COOs are becoming more immersed in technology decisions, while CIOs and CTOs—and IT staff members— are becoming part of high-level decision-making teams. Importantly, the CIO is becoming part of the leadership team of the composable enterprise. These leaders need to understand the technologies available to drive business forward, as well as the dynamics of their fast-changing marketplaces. These teams will be guided by the knowledge of IT executives combined with the acumen of business leaders.

- Composable enterprise leaders need to be proactive with the business. CIOs and IT leaders need to be proactive in engaging the business to understand its requirements. Forward-looking CIOs know that their role needs to involve more than responding to technology implementation requests from year to year. "It's not that marketing needs Salesforce, or human resources needs Workday," says Mason. "Enterprises need a strategy where they can bring these applications into the landscape and consume them rapidly. That requires a technology strategy, a process strategy of how those things hook into existing processes, or what new processes get created. If you don't fully understand what the business is asking for, you don't really understand what it needs."

CIOs are rapidly evolving into leaders, helping to guide corporate strategy as the composable enterprise becomes a reality. As chief innovation officers, CIOs have the role of identifying and putting in place the technology that will propel their organizations into leadership roles in the digital economy. Today's businesses need forward-thinking CIOs who will take charge and lead in developing composable enterprises. Otherwise, companies risk lurching into the digital economy rudderless and unable to compete effectively.

CHAPTER 4

The rise of the composable enterprise

by Joe McKendrick, Forbes staff writer

The new IT leader is going to be charting the direction of a new kind of company. Today's successful organization needs to be able to turn on a dime, changing its product or service strategy as fast as its customers' needs require. The successful business of the 21st century crosses all boundaries; can quickly meet and adapt to competition, whether it comes from another part of the world, another industry or a startup; or it can use its core competencies to extend itself in new ways. Welcome to the *Composable Enterprise*. This kind of company—powered by cloud, open APIs, data analytics, mobile and social, and connected to the Internet of Things—is redefining markets and raising consumer expectations. The composable enterprise casts away the hierarchical and hardwired systems and processes that defined its predecessors, and represents a radical rethinking of how technology can serve innovation and how innovation can serve customers.

Definition of the composable enterprise

The composable enterprise is a highly connected organization with business processes supported by on-demand services that are acquired and leveraged from the cloud and APIs, furnished by outside providers.

Any organization, regardless of what kind of business it's in, how many legacy systems it might have or how clunky its current processes are, can begin today to make the transition to a composable enterprise. It won't be an overnight transformation, but as changes happen, the positive effects of digital transformation will take hold. What is needed to begin this journey is an in-depth knowledge and understanding of the new systems and processes that are driving new businesses, how they interact and what they are capable of delivering. This transformation will deliver more than efficiency and lower-cost processing—it means heightened competitiveness through digital engagement and analytic decision making.

Our recent survey[5] of IT leaders confirms there are a growing number of companies transforming into composable enterprises—and it's not just startups in Silicon Valley. Established businesses are responding to this new digital environment by structuring themselves as composable enterprises, built out of connected software services, applications and devices. In addition, most are seeking to adopt the Internet of Things (IoT) and microservices technologies, further digitizing important aspects of their operations. These organizations are able to expose core competencies as digital assets that can be leveraged by others, thus extending the bounds of their enterprises. For example, Amazon has exposed its competency not only as an online retailer, but as an IT solutions provider through Amazon Web Services. Amazon repurposed its own hosting infrastructure as a sellable service in its own right, and now AWS is a billion-dollar business.

The movement toward the composable enterprise is driven by heightened competition and increasing consumer expectations. The survey finds that 86% of IT executives say they are under pressure to deliver services faster than last year. A majority, 72%, have an API strategy going forward, which is driven by the need to integrate new software with existing infrastructure, give business teams the ability to offer self-service IT, get more value from existing software and enable mobile applications. In addition, this new digital reality incorporates the emerging Internet of Things—approximately 75% of IT executives ranked IoT as important to their business plans over the next 12 months. This includes plans to integrate wearables, which are further empowering consumers to expect greater engagement and connectivity with companies.

---

[5] MuleSoft Connectivity Benchmark Report. July 2015.

Accordingly, within the last few years, the role of IT itself has been shifting dramatically—emerging from the confines of back offices and server rooms to that of high-level business strategists and even digital provocateurs, shepherding their businesses through the disruptions that are now turning entire industries upside down.

Many organizations face the need to compete in a global economy and serve new, emerging customer channels. "We were challenged by the inability of traditional IT to quickly respond to ever-changing business demands," relates an enterprise architect at one of the largest manufacturers of roofing materials in the United States. This includes being able to integrate the systems as well as leverage the big data now coming from all parts of the company's operations. The company has several strategic initiatives—all steeped in digital technologies—to meet these new demands, the enterprise architect says. "We have increased our speed to market. We have to be innovative. When we launch a new product line, we have to be nimble enough to take care of it."

In this new economy, "success depends on connecting the unconnected," says Ross Mason, founder and VP of product strategy of Mule-Soft. "To compete in today's environment, organizations need to connect applications, data and devices."

Digital transformation is forcing organizations to reconsider how to leverage their core competencies, and in some cases, even reevaluate their core competencies. Companies can now build ecosystems that extend their organizational boundaries by allowing others to incorporate those core competencies into their own products—such as the Amazon fulfillment model for third-party vendors.

At a more radical level, digital transformation enables reframing and redefining an organization's core competencies. An example is the New Zealand Post; in light of declining postal revenue from stamps, the company reframed its core competency from the physical delivery of postal mail to the provision of delivery information services such as address lookup, postal tracking calculators and tracking information, all delivered and monetized via APIs. The postal service established an internal team of digital "intrapreneurs" to explore and establish new channels to reach customers. "We were searching for new business models in what everyone sort of recognizes as a declining postal business model worldwide," says Joseph Brophy, solutions development manager for New Zealand Post Digital. "Our goal was to accelerate growth in e-commerce and parcels and logistics." Ultimately, Brophy adds, the goal of the initiative is to leverage digital approaches to enhance the customer experience and introduce design thinking to the nation's postal service.

What is a composable enterprise?

This composable enterprise is a lightweight entity, capable of quickly mapping technology solutions to ever-changing business requirements. The business is built on services that are acquired and leveraged from APIs provided by outside providers, or through a company's internal data centers. The services are connected through APIs, in the manner of building blocks. "The web is the playbook for the composable enterprise," Mason points out. "It taught us that you can decouple very complex systems with simple interfaces called APIs; as long as you make them usable, and you think about the consumer of those interfaces, they can drive enormous value. APIs have shown us that every business, no matter how different and complicated it is perceived to be, has technology components that can be broken down into smaller, composable pieces that can be consumed by the business."

The composable enterprise has the following characteristics:

- Business processes are assembled by APIs
- Organizational processes and data are opened up to partners and customers via APIs
- Business users can create front-end applications and access data, on demand
- The architecture and infrastructure is highly scalable, capable of accommodating new workloads and requirements on demand
- Technology solutions emerge not only from top-down vision, but from bottom-up capability
- Security is baked throughout the entire infrastructure, in every composable building block

This new breed of company moves quickly to adapt to changing customer realities and requirements, and is not dependent on any single type of technology. Therefore, the IT architecture behind composable enterprises is distributed and highly collaborative; central IT acknowledges and actively encourages a distributed architecture to encourage innovation, while still embedding governance and controls. In the past, IT was reactive: ordering and building systems to meet requirements, preparing and delivering reports, and troubleshooting systems issues. The new IT role is proactive, encouraging innovation and even acting as an agent of disruption for the business.

The rise of digital services—built on a foundation of APIs, virtualization and cloud—is changing the way organizations view applications and the way they handle enterprise IT. Organizations now can assemble their required functions from a range of both internal and external APIs—and quickly create new ones—that are capable of being integrated and retired on demand as needed by ever-shifting business requirements.

The services that form the composable enterprise cover a wide range of applications and functions—from sophisticated business and technical applications, to technology-level services such as storage and security, to mobile apps.

The development of composable enterprises will be enabled from the top down, but will be enriched by the activity and engagement of developers and business users at all levels of an organization. Within a composable enterprise, larger, more monolithic business processes are broken down and made available as more granular functions. As a result, it's possible to build and deploy applications and data at a much more rapid rate than within traditional settings, as many services and features can be pre-built, pre-tested and readily available to be pieced together, building-block style.

Such a new approach to technology calls for rethinking the way assets are deployed and made available. "We call this approach API-led connectivity," says Mason. "It is a different way to think about how to organize the assets that you own, the way you expose them and who you expose them to."

For technology leaders, the rise of digital, composable enterprises doesn't mean the deep-down technical tasks are going away—it means that along with providing traditional technical capabilities, IT departments also need to take a leadership role within the business, guiding their organizations through the fast-growing digital economy. It's no longer just about keeping the lights on, making sure the network runs and making sure one has the right software on the desktop— it means playing an active role in business strategy across all parts of the enterprise, from human resources to finance to production.

The composable enterprise is not limited to tech-savvy startups that can put everything they do in the cloud. Many established businesses, with legacy technology assets, are also evolving into composable enterprises. Organizations with legacy heritages as diverse as Amtrak, Unilever, News Corporation and the New Zealand Post are reinventing themselves as composable enterprises, with CIOs and technology leaders paving the way.

The people, process and technology behind composable enterprises

The composable enterprise is not just about technology, but rather is built and functions on three essential elements— people, process and technology. The technology fabric that binds this all together is essential for competing in a global economy and providing the information that decision makers need.

IT leaders and their departments have pivotal roles to play in this process—enabling the business to expose capabilities in new and innovative ways, such as designing and releasing digital products, and experimenting with new approaches. This process is about people and processes as much as it is about technology, Mason adds. It's up to "central IT to become both a partner and platform to the business, working to expose more of the value of the business."

Such a transformation has been taking place at New Zealand Post over the past five years, says Brophy. The move to API-based infrastructure has introduced "rationalization and simplification to our organization structure," he says. "From a technology standpoint, the pace has changed. Everything is getting faster and faster. A few years ago, some of what we were doing in technology would have been seen to be sufficient for the environment we were operating in. Now, it's not. That's why we have a digital team, with a focus on innovation. We know that we have to do things differently."

Make no mistake, the composable enterprise, from a technology perspective, is a demanding environment. Reliance on the legacy technology solutions that may have worked a decade ago—or even as little as five years ago—needs to be rethought. Many of the IT systems in existence today weren't originally designed to support composable enterprises built around APIs. Rather, they were built for an outdated vision of the future, according to the assumption that the business processes they support would also need to exist for years at a time, with occasional adjustments or upgrades along the way.

Building solutions to change is the key to the composable enterprise. All components—from services to applications to underlying processes—need to be flexible and agile, able to be changed or reconfigured on a moment's notice as business demands change.

The architecture of a composable enterprise must reflect this great adaptability. Such an architecture requires the following:

- Scalability on demand: Capable of calling services and functions as needed by the business.
- Adaptable to all devices and clients: Services and applications will run on any device the client is using at the time.
- Service-oriented architecture: Essential applications and functions are abstracted as scalable, reusable, loosely coupled services. Microservices ensure the granularity needed to adapt to key business functions.
- Automation: Manual processes or scripting need to be taken out of all aspects of the architecture so that on-demand services called by the business are readily available.
- Self-service: Business decision makers can access key components of applications and data without waiting for their IT departments.
- Accessible enterprise data layer: Key information needs to be made available to decision makers when they need it, from any source. Data sources, as they are identified, should be easily

and quickly integrated into the enterprise data flow in a standardized and repeatable way.

Ultimately, it means bringing IT operations and business operations onto the same page—a challenge that enterprises have been working to address since the dawn of the computer age. The composable enterprise is, in many ways, future-proof—it enables change and reconfiguration—even getting into entirely new lines of business. The right technology will always be available at the right time as organizations move into the future, and seek to compete in a hyper-competitive global economy. Having streamlined functional components that can be assembled, reassembled and disassembled on a moment's notice, as the business needs them, will go a long way in enhancing success in this new environment.

For News Corporation Australia, CIO Tom Quinn notes that the ability to leverage the architecture of the composable enterprise provides a way to acquire capabilities, on demand, without ripping and replacing existing investments. "We run a two-speed technology stack," says Quinn. "We have larger systems at the back end that are our systems of record, and give us the functions like CRM, HRIS and the financial system. At the front end, we have over 80 SaaS providers providing different pieces of the pie for digital delivery, covering ad serving, content augmentation and image enhancement."

The advantage of these SaaS-based applications is that they can be swapped on demand, Quinn says. "they are replaced often by a better mousetrap, and if we have a number of these systems that we can easily integrate together, we can take out the poor-performing player and replace it with something that works better for us."

Leading the composable enterprise

It takes innovation-minded, forward-thinking business leadership to transform from a staid, traditional enterprise to a composable enterprise, built by the business, for the business. "In order for us to be more agile and provide better visibility to our products, we have to engage our business," says the roofing company's enterprise architect. The path to the composable enterprise starts within the organization, engaging developers and employees at all levels—not just in the C-suite or in the data center. Often, the catalyst that first drives the movement to the composable enterprise begins with workarounds to IT departments, which may be not responsive enough to business requirements. The rise of cloud applications—commissioned by business users and developers outside of formal IT budgets—has become commonplace.

CIOs need to step up as Chief Innovation Officers and provide guidance and vision to help lead the composable enterprise. Increasingly, organizations that leverage technology in innovative ways are disrupting their industries and gaining market share. CIOs and their staffs will drive this evolution to composable enterprises, providing solutions—embodied as APIs.

The reliance on APIs for direct revenues, as well as enterprise-wide capabilities through integration—as indicated above—puts CIOs in business leadership positions. Strong leadership from the CIO is paving the way for New Zealand Post's transformation into the digital economy, Brophy observes. "Our digital work and objectives are highly visible at the executive leadership and board levels," he says. "Our API efforts have had very strong sponsorship from our senior leaders and our CIO."

Here are the key attributes of emerging composable enterprises, and the new roles CIOs—as Chief Innovation Officers—will play in helping to lead them:

IT needs to be a strategic partner to the business. All too often in organizations, IT has been the department of "no," or the place where project requests go to languish. In a composable enterprise, IT delivers capabilities, not projects. It serves as a full partner to the business, but at the same time, may even be invisible. Services rendered through the cloud, APIs and mobile are part of a structure that enables decision makers to quickly access the applications and information they need.

Understand and empower your IT consumers. Change how the business works with IT. CIOs and IT leaders need to be proactive in engaging the business in understanding their requirements. Forward-looking CIOs know that their role needs to involve more than responding to technology implementation requests from year to year. "It's not that marketing needs Salesforce, or human resources needs Workday," says Mason. "Enterprises need a strategy where they can bring these applications into the landscape and consume them rapidly. That requires a technology strategy, a process strategy of how those things hook into existing processes, or what new processes get created. If you don't fully understand what the business is asking for, you don't really understand what it needs."

The 'I' in IT means "Innovation" and "Integration." You now compete on how well you connect your assets to your audiences. The ability to connect requires the ability to rapidly add or provision services available from either within or outside the enterprise. "It's not just about connecting applications together, it's also about exposing information through things like APIs that enable people to be able to grab information easily, in a self-service way, without having to go through central IT for everything," says Mason. "This bolsters the reputation of IT of becoming a partner to the business."

Conclusion

Those enterprises that hang on to the old perception of information technology—as a data processing center somewhere in the back of the organization spitting out reports and providing system updates—are missing the opportunity of the century. "The challenge is changing your technology team's mindset from the old days when everything was good," says Quinn.

"Before, you could walk into a data center and touch your piece of hardware. Now, you can't. You don't know where pieces of software are running; they could be running anywhere in the world. The challenge is then trying to hook those different locations together for the end-to-end workflow of your business."

With the ability to rapidly build, publish and consume APIs, enterprises have a powerful tool at their disposal to move into the new digital business environment that will soon dominate the world's markets. The ability to leverage APIs and cloud services— versus buying or building software and hardware technology—provides enterprises with enormously flexible architectures. Those organizations that do not begin the journey to becoming a composable enterprise risk becoming bogged down in inflexible, legacy technology systems and processes and will be unable to rapidly respond to changes and challenges in their markets.

CHAPTER 5

Architecting the composable enterprise

by Jason Bloomberg, President, Intellyx

Introduction

The notion of the composable enterprise makes it clear that technology has infiltrated every corner of the business - and, quite frankly, there isn't that much difference between the business and IT anymore. Today's enterprises are software-driven organizations that shift the roles and perspectives of everyone in the organization, whether they be "business" or "technology."

The opposite of centralized IT, therefore, isn't only decentralized IT— although distributing the technology effort in the enterprise is an important aspect of this trend. It's important to decentralize how the business accesses and uses technology resources as well.

In order for a decentralized IT organization to take shape, IT must build out the capabilities the business needs while empowering business users to get their jobs done and thereby meet the needs of customers and the organization. In other words, IT has to shift from being the sole provider of technology to a strategic enabler of technology use throughout the business.

The end state of this combination of decentralization and empowerment is enabling organization by assembling and reassembling modular components—components made up of both technology and people.

Supporting this idea introduces profound business challenges for both business and technology leadership. The first step to tackle these challenges is to optimize the business' IT architecture for the composable enterprise.

The challenges of first-generation SOA

This notion of IT delivering composable capabilities to the business, as though application functionality and data were LEGO blocks for business people to assemble, is not a new idea. In fact, composability was an important benefit of Service-Oriented Architecture (SOA)—at least, the version of SOA IT shops struggled with during the 2000s.

SOA is an approach for abstracting enterprise software capabilities as reusable services in order to support more flexible business processes and ideally, more agile organizations. However, in retrospect, the original promise of SOA was largely unrealized.

Vendors used the approach to sell middleware, which led to expensive and difficult implementations. Deployments were centralized, leveraging hub-and-spoke technology that was ill-suited for the cloud. The SOA organization was centralized as well, with activities taking place internal to the organization, limiting the ability for teams at different organizations to share knowledge of lessons learned.

Eventually, the architectural focus on improving IT and organizational governance in order to achieve greater levels of business agility was largely subsumed into the technical minutiae of enterprise integration. The entire SOA exercise, therefore, became an exercise in tight coupling, connecting endpoints to endpoints.

The principles of SOA were solid, but most implementations were over engineered and too complicated. Web services proved to be too challenging. And worst of all, enterprises failed to emphasize the *consumers* of the services—both in the sense of the software endpoints that interacted with services, as well as the people who would want to consume service capabilities and compose them to solve business problems.

The rise of API-led connectivity

The first generation of SOA centered on the role of the Enterprise Service Bus (ESB), and sported Web Services as the primary type of service. However, among the main roadblocks preventing greater success with SOA were the complexity and technical limitations of Web Services. Even though these XML-based standards promised loose coupling, most Web Services deployments were nevertheless excessively tightly coupled, limiting flexibility overall.

The rise of cloud computing also shifted the SOA discussion. What cloud computing brought to the SOA table were the principles of horizontal scalability and elasticity, automated recovery from failure, eventually consistent data (or more precisely, tunable data consistency), and a handful of other now-familiar architectural principles. The cloud also emphasized the importance of decentralized computing.

The appearance of the cloud coincided with the exploding popularity of Representational State Transfer (REST). REST arose largely out of the ashes of Web Services, and helped organizations overcome many of the SOA roadblocks that had limited their success with the architecture.

As a result, second-generation SOA was REST-based and cloud-friendly, favoring lighter weight approaches to moving messages around than the heavyweight ESBs that gave SOA a bad rep. RESTful interfaces cleaned up a lot of the mess that Web Services left behind, as they were web-centric, lightweight, and far easier to use than Web Services.

The JavaScript Object Notation (JSON) also played an important role in the maturation of SOA, as it proved to be a simpler and more flexible data format than XML. The fact that JSON objects were themselves JavaScript also provided an ease of use that XML was ill-suited to deliver. Working with XML/SOAP is quite difficult; all modern JavaScript and mobile dev platforms now support JSON natively.

The combination of REST and JSON essentially moved the ball on application programming interfaces (APIs), leaving behind the challenges of Web

Services. Today, APIs are more likely to be RESTful, HTTP-based inter-faces than SOAP-based Web Services.

In fact, perhaps the most successful part of REST to date has been the simplification of the API. Enterprises no longer need a language- spe-cific protocol that depends upon sophisticated network controls under the covers. Today they can take HTTP for granted, and a simple request to a URL suffices to establish any interaction they care to implement between any two pieces of software, regardless of language.

APIs and the composable enterprise

The customer-centricity of digital efforts has led to the broader trend of the democratization of technology. No longer can enterprise apps afford to be immune to the pressures of consumer demands. Instead, everyone in the enterprise—or any other size organization, for that matter—expects the applications they use at work to be as convenient, flexible, and mobile-enabled as the apps they use anywhere else.

Enterprise IT, therefore, must respond to this rising tide of democrati-zation, not only by supporting mobile interfaces to enterprise apps, but also by empowering an entire ecosystem of digital capabilities for both internal and external users. Such users expect to find apps in app stores—marketplaces of functionality, originating both inside and out-side the organization.

Furthermore, users are expecting to compose the functionality of APIs and the data they provide to meet shifting business needs. Sales data should connect to business intelligence should connect to analysis and visualiza-tion—the list of such composition opportunities is endless. This expecta-tion of composability is at the heart of the composable enterprise.

From a technical perspective, the secret sauce that makes this app store-driven, user-centric composition vision come to life are the APIs that form the glue among the various application components and other services, including the small, modular components called microservices. And now that such APIs are web-centric, leveraging REST, JSON, and other easy-to-use protocols, there's no excuse for developers not to get them right.

Furthermore, there are so many available APIs that meet the needs and business models of so many organizations that an API economy has grown up around them. In this API economy, developers and other people can assemble apps from a mix of components built in- house and available in the cloud.

Companies who may have never thought of themselves as offering software-based products or services to their customers are now able to leverage APIs to expand their offerings. As enterprises in multiple industries become software-driven organizations, APIs become the means for providing value to customers, for maintaining efficient relationships with suppliers, and for participating in the broader commerce communities to which they belong.

New Zealand Post is a great example of such an organization. This company developed a number of APIs, empowering external users to integrate with its shipping, addressing, and postal systems. These APIs helped to build its parcel delivery business, which overtook mail delivery as the largest source of revenue in 2014. New Zealand Post is now using a commercial version of its addressing API to assist with identity verification on credit card applications.

Architectural leadership for the composable enterprise

APIs support enterprise composability at two levels. Inside the organization, people leverage APIs at the team level, while external to the organization, customers and partners can mix and match APIs from different places, composing and recomposing capabilities and information at will.

In some cases, the composability strategy also has to be driven from the top down. Executive-level concerns drive strategic initiatives that drive API-led composability. In other cases, composability is bottom-up, as developers put together capabilities to meet various project needs. However, in either case, the same APIs facilitate such composability – assuming, of course, they are architected properly to support such composability.

Architecture, in fact, is a critical enabler of the composable enterprise vision. APIs must be reusable and modular, and it goes without saying that nonfunctional requirements like compliance and security must be bulletproof. All of these requirements depend upon a lightweight, Web-centric architecture.

Where, then, should the IT organization go to get proper architectural leadership—or any other expertise necessary to execute on the composable enterprise? Many companies use the center of excellence approach.

Centers of excellence (CoEs) are teams and associated knowledge resources that provide leadership, best practices, research, and support for a focus area like architecture or APIs. On the surface, it sounds like such a center is just the ticket to help an organization transition to becoming a composable enterprise.

There's just one problem with this plan; CoEs by definition centralize expertise. They actually become roadblocks for IT, as well as for line of business people who want access to IT capabilities and data, because leveraging a center of excellence requires formal requests to a small, typically overworked team. In contrast, the decentralized alternative to a CoE is a *center for enablement*. Centers for enablement focus on rolling out new capabilities and assets to a skilled audience – often with the support of a self-service capability like a portal or app store.

Instead of acting as an ivory tower repository of expertise, a center for enablement distributes templates that various audiences can use as starting points. The goal is to teach people 'how to fish' as well as '*where* to fish' for the expertise and capabilities they require.

Templates, APIs, app stores, and marketplaces all facilitate self- service access to IT resources. Moving to a self-service model requires changing behavior in the organization. Get this model right, however, and it frees IT to focus on more important tasks.

IT for the 21st century: Supporting the composable enterprise

The vision of the composable enterprise is a vision of a world where IT is a partner to the business. IT is no longer a back office function, but rather an enabler of the business. To make this vision a reality, the approach cannot only be top-down or bottom-up. It must be customer-driven and end-to-end—the essence of digital transformation.

At one end, of course, are legacy systems, which will continue to present a challenge to the composable enterprise. However, rip and replace is rarely the best option. Instead, expose legacy assets as APIs, and modernize them how and when delivers the most value to the organization and its customers.

APIs make even inflexible legacy assets composable as part of the API economy. APIs are the linchpin of a modern approach to integration that will create flexibility, agility, efficiency, and ultimately business success for years to come.

The business no longer has the time to wait for centralized IT. But that doesn't mean the IT organization goes away. Instead, there are the three roles that remain critically important for modern IT: security, governance, and maintaining access to systems of record—via APIs.

The role of architecture is shifting as well. The top priority for architecture is supporting the organization's business agility goals—helping the organization deal with the change at the heart of the composable enterprise.

In order to successfully become composable enterprises, organizations must decentralize the IT organization and support centers of enablement rather than centers of excellence. Implement lightweight, web-centric architectures and the design principles of API-led connectivity. Organizations must focus on teaching the whole organization how to fish—they must implement lightweight, web-centric architectures and the design principles of API-led connectivity, deliver great user consumer experience of the reusable assets and create a center for excellence to keep teaching consumers how to discover and use assets.

Such business and technological transformation is difficult, but the urgency and path to success is clear. Every organization, in any industry, has the power to become a composable enterprise.

CHAPTER 6

The next step in the evolution of SOA: API-led connectivity

by David Chao, MuleSoft

As you've seen, organizations must adopt a new IT operating model to adapt to the increasingly demanding market. But how do you connect the composable parts of your enterprise? Existing connectivity approaches are not fit for these new challenges. Point-to-point application integration is brittle and expensive to maintain. Service-oriented Architecture (SOA) approaches provide some instruction in theory, but have been poorly implemented in practice. The principles of SOA are sound - well-defined services that are easily discoverable and easily re-usable. In practice, however, these goals were rarely achieved. The desire for well-defined interfaces resulted in top-down, big bang initiatives that were mired in process. Too little thought, if any, was given to discovery and consumption of services. And using SOAP-based Web Services technology to implement SOA proved to be a heavyweight approach that was ill-suited then and even more ill-suited now for today's mobile use cases.

In addition, as previously mentioned, many people within and outside of an enterprise need to be able to access parts of the network using consumption models that are familiar to their way of working. IT cannot satisfy the demands of the business by being the sole technology provider for the organization; through the application network, described in Chapter Two, the organization can achieve *reuse* of services, knowledge, access and best practices, which provides much more technology capacity to the business. Our vision of an application network is realized through our approach to modern enterprise integration, which we call API-led connectivity.

API-led connectivity: The evolution of SOA

While connectivity demands have changed, the central tenets of SOA have not, that is, the distillation of software into services that are well-defined, reusable, and discoverable.

This vision is perhaps even more important given the proliferation of endpoints. The complexity of providing multiple stakeholders customized views of the same underlying data source, whether it be a core banking system or an ERP system, increases exponentially with the number of channels through which that data must be provided. It also reinforces the need for data at the point of consumption to be decoupled and independent from the source data in the system of record, becoming variously more coarse-grained or fine-grained as the use case requires.

This problem lends itself to a service-oriented approach in which application logic is broken down into individual services, and then reused across multiple channels. Yet the heavyweight, top-down implementation approaches previously noted are not a fit for the agility that today's digital transformation initiatives demand.

To meet today's needs we propose a new construct, API-led connectivity, that builds on the central tenets of SOA, yet re-imagines its implementation for today's unique challenges. API-led connectivity is an approach that defines methods for connecting and exposing your assets. The approach shifts the way IT operates and promotes decentralised access to data and capabilities while not compromising on governance. This is a journey that changes the IT operating model, enables the application network, and which realizes the goal of enables the realization of the "composable enterprise", an enterprise in which its assets and services can be leveraged independent of geographic or technical boundaries.

API-led connectivity calls for a distinct "connectivity building block" that encapsulates three distinct components:

- Interface: Presentation of data in a governed and secured form via an API.

- Orchestration: Application of logic to that data, such as transformation and enrichment.

- Connectivity: Access to source data, whether from physical systems, or from external services.

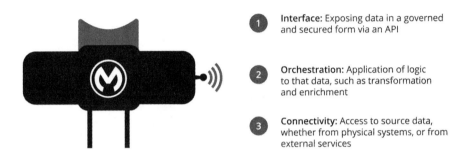

1 Interface: Exposing data in a governed and secured form via an API

2 Orchestration: Application of logic to that data, such as transformation and enrichment

3 Connectivity: Access to source data, whether from physical systems, or from external services

Designing with the consumption of data top of mind, APIs are the instruments that provide both a consumable and controlled means of accessing connectivity. They serve as a contract between the consumer of data and the provider of that data that acts as both a point of demarcation and a point of abstraction, thus decoupling the two parties and allowing both to work independently of one another (as long as they continue to be bound by the API contract). Finally, APIs also play an important governance role in securing and managing access to that connectivity.

However, the integration application must be more than just an API; the API can only serve as a presentation layer on top of a set of orchestration and connectivity flows. This orchestration and connectivity is critical. Without it, API to API connectivity is simply another means of building out point-to-point integration.

API vs. API-led connectivity

Stripe, as an "API as a company" disintermediating the payments space, is an archetype of the API economy. Yet at MuleSoft's 2014 CONNECT conference, Stripe's CEO John Collison was quoted saying "you don't slather an API on a product like butter on toast." Thought of in isolation, the API is only a shim that, while hiding the complexities of back-end orchestration and connectivity, does nothing to address those issues. Connectivity is a multi-faceted problem across data access, orchestration and presentation, and the right solution must consider this problem holistically rather than in a piecemeal fashion. To only consider APIs is to only solve only one part of the connectivity challenge.

"Three-layered" API-led connectivity architecture

Large enterprises have complex, interwoven connectivity needs that require multiple API-led connectivity building blocks. In this context, putting in a framework for ordering and structuring these building blocks is crucial. Agility and flexibility can only come from a multi-tier architecture containing three distinct layers:

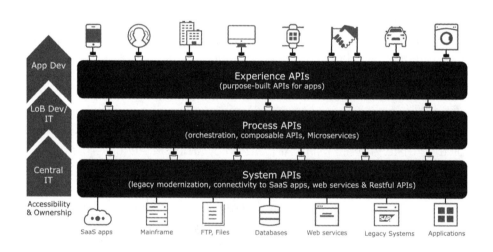

- System Layer: Underlying all IT architectures are core systems of record (e.g. one's ERP, key customer and billing systems, proprietary databases etc). Often these systems are not easily accessible due to connectivity concerns and APIs provide a means of hiding that complexity from the user. System APIs provide a means of accessing underlying systems of record and exposing that data, often in a canonical format, while providing downstream insulation from any interface changes or rationalization of those systems. These APIs will also change more infrequently and will be governed by Central IT given the importance of the underlying systems.

- Process Layer: The underlying business processes that interact and shape this data should be strictly encapsulated independent of the source systems from which that data originates, as well as the target channels through which that data is to be delivered. For example, in a purchase order process, there is some logic that is common across products, geographies and retail channels that can and should be distilled into a single service that can then be called by product-,geography- or channel-specific parent services. These APIs perform specific functions and provide access to non-central data and may be built by either central IT or line of business IT.

- Experience Layer: Data is now consumed across a broad set of channels, each of which want access to the same data but in a variety of different forms. For example, a retail branch POS system, e-commerce site, and mobile shopping application may all want to access the same customer information fields, but each will require that information in very different formats. Experience APIs are the means by which data can be reconfigured so that it is most easily consumed by its intended audience, all from a common data source, rather than setting up separate point-to-point integrations for each channel.

Each API-led connectivity layer provides context regarding function and ownership

Layer	Ownership	Frequency of Changes
System Layer	Central IT	6-12 months
Process Layer	Central IT and Line of Business IT	3-6 months
Experience Layer	Line of Business IT and Application Developers	4-8 weeks; more frequently for more mature companies

Benefits of API-led connectivity

The benefits of thinking about connectivity in this way include:

Business

- IT as a platform for the business: By exposing data assets as a services to a broader audience, IT can start to become a platform that allows lines of business to self-serve.

- Increase developer productivity through reuse: Realizing an API-led connectivity approach is consistent with a service oriented approach whereby logic is distilled to its constituent parts and re-used across different applications. This prevents duplication of effort and allows developers to build on each other's efforts.

- More predictable change: By ensuring a modularization of integration logic, and by ensuring a logical separation between modules, IT leaders are able to better estimate and ensure delivery against changes to code. This architecture negates the nightmare scenario of a small database field change having significant downstream impact, and requiring extensive regression testing.

Technical

- Distributed and tailored approach: An API-led connectivity approach recognizes that there is not a one-size-fits-all architecture. This allows connectivity to be addressed in small pieces and for that capability to be exposed through the API or microservice.

- Greater agility through loose coupling of systems: Within an organization's IT architecture, there are different levels of governance that are appropriate. The so-called bi-modal integration or two-speed IT approach makes this dichotomy explicit: the need to carefully manage and gate changes to core systems of record (e.g. annual schema changes to core ERP systems) while retaining the flexibility to iterate quickly for user facing edge systems such as web and mobile applications where continuous innovation and rapid time to market are critical. Separate API tiers allow a different level of governance and control to exist at each layer, making possible simultaneous loose-tight coupling.

- Deeper operational visibility: Approaching connectivity holistically in this way allows greater operational insight that goes beyond whether an API or a particular interface is working or not, but provides end-to- end insight from receipt of the initial API request call to fulfilment of that request based on an underlying database query. At each step, fine grained analysis is possible, that can not be easily realized when considering connectivity in a piecemeal fashion.

What the layers do is release tension in the organization. The system API layer is all about unlocking assets and decentralizing access. If you start here, you start making that data available to you, so the next time somebody asks you for reporting information or cataloging information, do it with an API mindset. Open up an API and give the API to them. Many organizations have already invested in web Services or full-blown SOA. This is a great starting point towards API-led connectivity and those services can be leveraged. The critical point (which is sometimes missed) is the role of the API is to make those services more consumable by other people in the business, not just the SOA architects who are trying to deliver projects faster.

The second layer is actually all about driving agility and new value creation. What you're actually saying is to the business, "Go ahead. Use the data. You have access to it. We govern it. We'll make sure that you do the right things with the data, but you can go and self-serve. You can go and build new value on top of the data that we have in the datacenter in our applications."

The top level is all about innovation. It's about driving new types of application experiences that obviously gives you happy customers.

This is how our customers are getting themselves plugged into the wider API economy. Wells Fargo is a good example; they are creating a set of platform APIs to allow retailers and app developers to leverage things like location services. Having location-based services means that they can start to engage with their customer's customer in a different way and to drive new experiences further out more like a B to B to C model. These experiences can be different for different type of applications. It's not just about Wells Fargo owning the whole end-to-end chain. They are plugging in their value and allowing other people to build value on top of it.

The customer journey to API-led connectivity

Realizing an API-led connectivity vision must be much more than a technology decision. It requires a gradual but fundamental shift in IT organizations' architectural vision, development approach and the way developers approach their roles. The challenge is one as much about process change as it is about technology implementation.

Application networks built through API-led connectivity are designed to enable the whole org, not just IT, and provide a distributed model while also ensuring governance and control. This shift in how IT organizations envision themselves has had powerful results among companies who have embarked on this journey.

Case study: Top 5 global pharmaceutical company

A pharmaceutical company's key route to market is its relationship with its physicians. Physicians have a choice of drugs to prescribe, and the timeliness and richness of drug information is a critical determinant of which drug the physician prescribes.

This global pharmaceutical company's marketing team set out to transform how the company engaged its physicians by refreshing its web presence, and developing a new mobile channel, in all 140 of the countries in which it operates. Prior to the introduction of this project, this company suffered from a mix of different technologies in each country with inconsistent design and execution. The challenge was to drive greater control and governance, but at the same time, give each country the flexibility and freedom to move quickly and drive in-country localization where appropriate.

To realize this vision, this company embraced the API-led connectivity approach. It considered the web and mobile experience that it wanted users to have and then built an experience API layer that spanned across web and mobile channels.

IT then built a set of process layer that exposed key pieces of logic which could be leveraged by each in-country IT team. For example, identity management is a key requirement of such a digital marketing platform. Each time a doctor logs onto the platform, a real-time check must be made to ensure that that doctor is licensed to practice medicine. The system of record for such information varies by country, and within each country, by clinical specialty. MuleSoft's Anypoint Platform sits as the orchestration layer receiving, routing and transforming identity management requests in real time.

However, realizing the API-led connectivity vision is not a discrete goal, but rather a continuous journey. Moreover, it is a goal that can be only be realized in incremental steps. Through partnering with dozens of Fortune 500 companies on their API-led connectivity digital transformation journeys, we have distilled best practice into the following steps:

- Start-up mode: In order for the API-led connectivity vision to be successful, it must be realized across an organization. However, in large enterprises it is simply not possible to wipe the slate clean and start from scratch. Consequently, the API-led connectivity customer journey must start with a vertical slice of the business, for a specific use case or for a specific line of business. By bounding the problem, the scope of change is reduced and the probability of success increased. Training and coaching to drive role modeling of new behaviours is critical at this stage.

- Scale the platform: Once initial proof points have been established, these use cases will naturally become lightning rods within the organization that will build mindshare and become a platform to leverage greater adoption. In addition, the service oriented approach results in the natural creation of reusable assets which exponentially increases the value of the frame-work as the number of assets increases.

- Build a center for enablement (C4E): Once scale has been established, it's critical to quickly codify best practice and pro-vide a platform for discovery and dissemination through the organization. The result of such a process is mass adoption across the enterprise. The core of this C4E may also be built during the startup mode and scaled as required. The C4E will be discussed further in the next chapters.

Case study: Top 5 global bank

Digital transformation is often considered to be an external phenome-non. However, whether in terms of enabling transformation outside the company, or in and of itself, digital transformation is a powerful phe-nomenon *inside* organizational boundaries also.

This multinational financial services company wanted to drive a firm wide architecture driving application development consistent with one of six best practice messaging patterns

This approach has initially seeded into one line of business. This suc-cess prompted subsequent rollout across 13 lines of business globally, connecting more than 1,000 applications in production.

In the initial startup mode, the enterprise seeded adoption via a central group which was better able to seed adoption and prove out the approach. As the company continues to scale across the business how-ever, it is looking towards API-led connectivity as the means to decen-tralize elements of the architecture to drive scale, yet maintain control.

Central to the ability to realize this vision was a center of excellence which helped to codify knowledge and disseminate best practice. Mule-Soft helped to build out this center for enablement through delivering on its proven customer journey approach.

How to implement API-led connectivity in your organization

The application network, built through API-led connectivity, isn't just a technological quick fix, it's a new mode of operating, and as such, requires a number of changes. We recommend the following approaches to get started:

- Adopt an API-led connectivity approach that packages underlying connectivity and orchestration services as easily discoverable and reusable building blocks, exposed by APIs.
- Structure these building blocks across distinct systems, process and experience layers, to achieve both greater organizational agility and greater control.
- Drive technology change holistically across people, processes and systems in an incremental fashion.

The last point is part of the strategic leadership role that the CIO must take on in the composable enterprise - driving change and a new way of seeing technology implementation throughout the enterprise. In the next chapter, we'll take an in-depth look on how to do this.

CHAPTER 7

Creating a climate for change across the business

by Ross Mason, Nilanga Fernando,
Matteo Parrini and Guy Murphy, MuleSoft

Every part of an enterprise needs technology to build new applications for their specific function or customer. IT needs to transform from its traditional role as the sole technology provider to become an adaptive, responsive, and nimble organization that can keep up with the pace of the digital era as well as embrace the opportunities provided by a change-driven environment. This transformation can occur only if IT transforms itself into a strategic business enabler rather than a centralized technology provider.

Being an enabler means that IT has to decentralize and democratize application development and data access to the different lines of business (LoBs) and functional business partners. This way, IT can concentrate on a partnership with the business—i.e. providing a set of strategic and consistent assets and technology. As we've discovered with the concept of an application network, IT teams need to shift to providing services, knowledge, and best practices that can be used by the entire network.

In order to make that shift, an organization needs to both adopt enabling technology and build up new operational practices around a powerful network. These practices will need to enable the business to have selfserve access to IT development capabilities and data assets without compromising IT and data governance. Self-service means that the IT services and capabilities can be easily and readily consumed and therefore much of the underlying complexity of IT is masked from the business community who are often not IT experts.

In order to balance the creation of reusable assets, provide federated access to integration, enable self-service and mask complexity, a new organizational capability is required. We call this a *center for enablement*. This is a cultural shift away from IT delivering projects to the business to IT delivering reusable assets to enable the business to deliver their own projects the right way.

In this overall context, a center for enablement (C4E) is defined as: An organization in charge of enabling business divisions to successfully fulfill specific connectivity needs. It is responsible for providing a framework and set of assets to allow both the business and IT to build, innovate and deliver over their objectives in an agile and governed way. This operating model provides decentralized access to the most critical IT resources enabling innovation, reusability and scalability in an application network environment. The long term aspiration is for the C4E to enable LoBs, as well as IT, to consume *and* create these reusable assets.

What are the top three outcomes for a center for enablement?

In today's environment, many businesses pursue three top-level IT strategy priorities:
1. A cost-effective and agile IT landscape;
2. an improved relationship with the business;
3. the creation of digital and mobile strategies.

All of these require a step change in how the organization scales agility and improves time to market for enabling the business, while also consolidating the technology estate onto fewer, more futureproofed business platforms that will be organized primarily by business capabilities. Hand in hand with this philosophy is the desire to simplify the processes of engagement between the business and IT.

Against this backdrop, we have identified three key overarching outcomes the C4E needs to achieve for the enterprise as a whole:
1. Shifting the role, and culture, of central IT from project delivery to a more agile enabler for the business by creating reusable IT capabilities and assets and by demonstrating the

value of using those capabilities to drive change.

2. Improving the 'time to market' for how quickly the business and IT can respond to threats and opportunities by building capabilities that can be leveraged directly and indirectly by the different lines of business.

3. Establishing a core set of reusable assets that will create the composable enterprise, enabling different delivery capabilities within the business, including developer groups, Sis, and ISVs.

The main principle of a C4E

The main principle of a center for enablement (C4E) is to make it possible for enterprises to achieve two seemingly opposing goals of software design: governance and high delivery speeds. This is often referred to as "bimodal IT," but also enables the emerging multispeed IT that acknowledges the reality that different projects require different approaches for creation and delivery. The aim is to achieve a balance between risk and agility, so that areas that need to go at a faster pace can do so. For this to happen, enterprises need to support agile practices with right-sized governance underpinned by technology.

IT must no longer be perceived as a centralized capability delivering all technology related projects; furthermore, IT should provide capabilities to the business as a whole. With business self-service, however, comes the risk that the landscape, systems and processes can become once again highly fragmented and difficult to manage. There still needs to be "just enough" governance, security, and data access control that protect the enterprise's digital assets. For example, legal and regulatory requirements must always be met, no matter what type of business initiative is being delivered.

For this to work, the mindset of the organization needs to evolve to think about IT project delivery in a different way. Later in the chapter we will contrast the differences between traditional low-speed IT delivery and high-speed IT delivery. This can also be thought of as a high level framework to determine which projects should flow through a hybrid IT approach.

The cultural change is gradual and is supported by a C4E enabling audiences outside of central IT to self serve assets such as coding templates, APIs, examples and education materials in order to fulfill their needs without being dependent on a central function for all IT needs. This serves the newer breed of application developers that often sit outside the walls of central IT - either in the line of business IT or in the wider business itself.

For example, it is common for companies going through this change to adopt SaaS platforms like Salesforce.com. Typically Salesforce.com will need to connect to other SaaS and on-premises applications to enable its use with integration into existing workflows and employee applications. To support this need, the C4E would define integration templates for connecting applications such as SAP and Salesforce. These templates would capture common scenarios requested by the business and would enable teams (either centralized or decentralized) to make connections between applications through configuration instead of coding. Furthermore the template would contain the necessary application management, logging, security and error handling that complies with IT security and governance. The difference from prior approaches is that these templates are designed to be used by other developers outside of central IT. It also means that the discovery and self-serve aspects of these templates is intentionally designed through the C4E.

Another example of a reusable asset is an API for accessing product information in SAP (and/or other systems). An API provided by the C4E provides a self-serve developer portal to gain access to the API and make calls against it. The API is governed and managed through policies that the C4E/Central IT set and evolve over time - this may also include the Lines of Business taking ownership of existing APIs and building new ones, with governance and policies established by the C4E, when their knowledge and capability has matured.

The value of access through the API is that IT can see which applications are using the data, how frequently it's being used and throttle traffic flow if the API is being accessed a lot. Furthermore, the API can be incorporated into an integration template to connect the data to another system, such as an analytics platform. This means a business analyst could consume the data from the API into a data warehouse or reporting engine without having to write code.

The scope of a center for enablement

What a C4E does

The goal for the C4E is to decentralize the capabilities of IT without compromising its governance functions. It does this by:

1. Providing reusable assets that can be consumed by audiences outside of IT to deliver solutions that usually IT would be asked to deliver and would have delivered with a high-speed IT project mindset.
2. Reducing the delivery time for these solutions because the templates and APIs do a lot of the heavy lifting, allowing LoB IT or a business analyst to configure the templates without any need for coding.
3. Adhering to security and governance because these templates and APIs have security and governance built in. Therefore, these solutions don't create downstream problems and don't encumber the consumer with complicated details of how the security and governance works.
4. Reducing the pressure on IT to deliver every technology project, freeing up their time to work on more strategic projects.
5. Reducing the time from "idea to delivery", by providing reusable components and a set of guardrails that minimize complexity and debate within projects.

The C4E has to adopt a pragmatic approach in terms of content delivery, rather than be documentation-oriented it should be asset-oriented - where an asset is any kind of implemented example, API with a developer portal, already pre-configured component, etc. This is critical for the success of the C4E and is one of the key differences from a traditional Integration Center of Excellence (CoE).

Documentation will still remain as part of the content delivered by the C4E, but the approach needs to be centered around the heavy use of any assets produced, rather than the guidance to inform that use.

An alternative way to to consider the scope of the C4E is along the following parameters:

Reusable assets	These are defined as assets that have been designed and built across the enterprise, for use elsewhere - e.g. APIs, Templates, Connectors, Project Accelerators, or SDKs.
Core processes	The C4E's core processes will be focused on engagement (stakeholders), enablement (assets and capabilities), and optimization (of assets, cycle times, governance).
Organization	The C4E will require an organizational construct that provides a clear view of sponsorship, ownership, roles, responsibilities, and key capabilities and skills.
Governance	The specific governance structure around the C4E will require detailed thought. It's important to start with understanding current governance structures that could be considered either as constraints or as models of good agile practice.

How does a center for enablement differ from a center of excellence?

There are some perceived similarities between the Center of Excellence (CoE) strategy that many enterprises are familiar with and the C4E concept. However, the goal, focus, speed and audience of the two concepts are different. This is not to say that the role of the CoE is no longer required - this will continue to be a core competency for IT. Instead, the changing nature of consumer and business expectations requires a C4E to enable both IT, and developers outside of core IT, to be able to leverage and build on core IP using simpler reusable building blocks (such as APIs and templates).This should allow for more rapid delivery of projects and a wider group of people being able to deliver projects that adhere to core standards and governances rules, as governance should be built into the building blocks themselves. The following table highlights the key differences:

Delivery model	Project focused, centralized within IT or defined by Central IT	IT builds for capabilities. Enables the LoBs to deliver some projects independently
Governance & access	Strict and rigid, limited access to IT systems	Governed through APIs, self-serve access
Assets	EA patterns, domain architectures, best practices, security and access control models	APIs, domain architectures, templates and project accelerators
Consumer (of the assets)	Central IT, SIs	IT, SIs, LoB IT, App Dev. Creative, ISVs
Consumption model	Centralized, request-based (through projects)	Developer portals, data marketplaces, SDKs, embedded tools
Ownership	Central IT owns the infrastructure, applications and data	Central IT is a steward of the platform, access is self-serve
Enablement	Complex document driven	Evangelism, community building, self-pace training
Domain expertise	Centralized	Decentralized / Federated

Establishing a C4E

A C4E builds reusable assets to be consumed by other parts of the organization, not just IT.

One of the main challenges to be overcome as quickly as possible is to define the first set of assets and content that will be published in the C4E store. These need to trigger the transformation and also meet the critical requirements dictated by business strategy and the most critical projects. A successful approach must keep in consideration not only the most critical business initiatives, but also the relatively new capabilities just established in the C4E.

Typically, initial stages start with Central IT, then LoB IT and some core reusable assets. Phase Two should extend these self-serve capabilities further into the rest of the business. The federated C4E should build on the foundational assets with those relevant to their applications, data and device landscape. For example, a digital employee could adopt Salesforce as a core platform and would require a specific set of assets or capabilities to connect this to other systems or processes. Central IT may not even be involved in delivering these assets. For this to happen, these assets need to be assembled from core assets, created by the C4E that have the governance and security already built in (i.e. API templates or a base template for flows). Key to success in phase Two will be a light approach to evaluate the performance, adoption, suitability and reusability of assets - regardless of where they are created.

C4E consumers

The goal for the C4E is to enable new audiences within an organization to leverage assets and IP built by the company. This enables these consumers to self-serve key capabilities, data and services without reinventing the wheel each time.

The C4E needs to understand the motivations for these different audiences and build a strategy to provide them the key capabilities, data, and services they need to be able to contribute and/or build their own projects.

Role	Description / Skill	Typical Responsibilities
Core asset developers	These are specialist integration developers with deep integration skills and are deeply familiar with data transformation and legacy modernization. Focused primarily on system and process APIs, complex integration patterns, custom connectors.	Unlock data assets and initiate population of the C4E asset base. Act as quality entry point for any incoming assets into the C4E. Most likely within the C4E.

Role	Description / Skill	Typical Responsibilities
Application developers across the business	Developers with a primary focus on application experience (and may have some integration skills). They require flexibility and will exercise their own choices on how to integrate and which tools to use. Focused primarily on experience APIs.	Meet the business requirements for their initiatives quickly and in the most effective, engaging way. Drive demand and feedback into the C4E.
Integration developers across the business	These developers have basic to intermediate integration skills, and a good understanding of integration requirements driven by their project. Focused primarily on process APIs, complex process patterns such as business event and error handling.	Most likely part of the C4E community, whether in Central IT or across the business Meet business requirements for their initiatives. Provide a key input into the C4E for demand generation, as well as create reusable integration assets for other consumers.
Application administrators across the business	These administrators are focused on managing, monitoring, and operating applications for the company—including looking at performance, and access rights.	Their primary responsibility is to keep the application up and running.

Role	Description / Skill	Typical Responsibilities
Business analysts	They are consumers of data coming from various sources of data—legacy applications, SaaS applications, and other custom apps.	They interpret data and provide insight for the business.
Product owners	These are specialists within the LOBs who are thinking about new applications and thinking about adding new capabilities to current applications.	They set the business agenda or provide direction to application developers on what the use case is that needs to be implemented.

The C4E organization, structure and key roles

If your business has just started its journey to establish a C4E, it needs to develop and govern all the ancillary activities around the setup and operation of this new critical team. This new organization has to be integrated with the wider landscape of other teams, departments and processes - leveraging possible synergies and spreading innovation to infuse the transformation across the company.

One key aspect of C4E is a clear recognition that expertise will be developed in the federated teams, to both support this and maximize the value of this capability. Federation of integration disciplines and capability requires:

- Collaborative relationship: The development teams cannot afford to operate completely independently. Each team will function most effectively when guided and enabled by a form of integration competency team (CoE) and facilitated capability (C4E), as a way of recognizing opportunities, delivering shared services and working with each other.
- Clear language: The different competency teams must set up a common vocabulary. For example, if the SOA team uses the

term "service implementation" to refer to a set of components that implement a service, the LoB and mobile teams should use the same term.

- Best-practice enablement: One team with best practices must recognize the value of roles in other teams and must incorporate the roles into its best practices. Incorporating roles from other disciplines brings best-practice process knowledge from other disciplines and creates an opportunity to link processes where appropriate.
- Technology standards: The different teams and organizations (C-IT, LoB, mobile) typically deploy products with overlapping capabilities, where possible tools, and information access repositories should be standardized where possible. It's crucial to establish standards and development approaches to support the different requirements.
- Tooling integration: A strong focus on shared models, assets and IP will be of key importance to enabling seamless integration and semantic consistency of various tool types (toward the ideal of leveraging them together toward a common purpose), as well as the agility in shifting between integration patterns.
- Organizational structure: We do not intend to define the specific organizational construct required by the business here. This structure is something that needs to be designed with considerable specificity and understanding of the particular organization. Organizational design will need to take into account the following factors to influence the shape, level of federation and overall size of the new organization:
- Key project demands: These demands will drive the first asset creation and engagement with the C4E
- An understanding of current skillsets and capabilities within the company
- Long term connectivity priorities and strategies at a business level
- Alignment with a DevOps model and implementation

Below are listed some of the key future roles in a C4E - either as 'suppliers', 'consumers' or both. The table below is not intended to be exhaustive and defines only a set of roles and key responsibilities that we think should be part of the C4E:

Role	Description / Skill	Typical Responsibilities
C4E developers	Deep experience of integration development, API and Agile methodology	Develop C4E assets
C4E analysts / analysts	Integration and business analysts who are able to understand project-level integration requirements and translate these to C4E assets (and vice versa)	Generate appropriate demand for the C4E by triaging project requirements into priority self-serve candidates
C4E coaches & evangelists	Experienced in evangelizing, provoking and coaching teams	Evangelize and coach teams across the business to think differently and adopt the C4E's ways of working
API product owners / C4E asset owners	From Central IT or the rest of the business - the API / asset product owner should understand and apply product management fundamentals to each API or asset	Champion the API, engage the rest of business to reach mass adoption. Keep API operational and optimize through API lifecycle (inception through to deprecation)
C4E architect(s)	Deep experience of integration and API architecture - and a thorough understanding of industry trends, vendors, frameworks and practices	Provide "enough" governance over the design and operation of the actual C4E assets. These architects could be part of a wider community, and not just sit within the C4E

Role	Description / Skill	Typical Responsibilities
C4E lead / sponsor(s)	Owner of the C4E within the business. Strategic and operational management and leadership experience	Manage the overall success of the C4E, manage operation on a daily basis, measure ROI and performance, manage senior stakeholder and management perception, manage budget and funding

Key principles for building the C4E

Detailed planning for the execution of the C4E requires the previous aspects on scope, organization, audiences, and enablement to be agreed and confirmed. We have found that there are some common principles against which more detailed planning can take place:

- Unlocking core assets: Opening up and making available key existing data, leveraging capabilities and requirements coming from priority projects once data quality has been tested for and ensured.
- Business-led or application design driven: Unlocking data needs to start with the definition of priority projects / applications that need to be built which translates into the definition of the core APIs. This needs to be done in light of a wider business-facing consultation to ensure alignment with business priorities
- Fast iterations with collaboration with the application owners: Providing assets on the basis that feedback and improvement is expected and required. Consumers will rather get their hands on APIs that fulfil 80% of their requirements in 20% of the time, than 100% of their requirements over a longer time-frame. Requirements also change over time.
- Prioritizing customer APIs where there's significant traction: If you have a specific set of APIs that are gaining traction across the organization, such as surfacing customer data from

salesforce.com—these could be a suitable focus for the development of specific APIs for specific consumers across the business.

The outcomes from the composable enterprise can't merely be met through a change in enterprise architecture; they have to be accompanied by a change in business strategy and philosophy on how the entire business—not just IT operations—interacts with technology. This philosophical shift from a technology provider to a strategic business enabler is key for a business to truly realize the opportunities that the composable enterprise can provide – e.g. the flexibility, agility, and speed a modern business needs. The Center for Enablement represents a model of how to reorganize IT along these lines of thinking, and is a pathway to successfully transform IT for the change-driven enterprise.

Part II

Strategic initiatives that could help your business

We've tackled the big-picture elements of the enterprise in the digital age: the changing role of the CIO; IT's role in the modern business; creating a composable enterprise, in which experiences are created and recreated through small, self-contained services; and creating an environment where everyone in the business feels enabled and empowered to use those services to achieve business objectives.

But there are a number of specific, strategic initiatives that companies need to think about in order to stay competitive in today's business environment. The following chapters will address those initiatives in more detail, including primers on API strategy and security, the cultural shifts created by microservices and DevOps, and what IoT and mobile devices can mean for the modern organization.

One thing the subsequent chapters will make very clear; the role played by the CIO is not just about making adjustments to a technology stack or implementing some cloud software. We are recommending new ways of thinking about developing APIs, providing extraordinary experiences for your customers and making it easy for your developers to deliver products on a continuous basis. It is a rethink of the entire IT culture.

CHAPTER 8

Setting up an API strategy

by Julie Craig, Enterprise Management Associates
Research Director, and Reza Shafii, MuleSoft

Today everything truly is connected to everything, and APIs are the con-nection mechanisms. Many companies are finding that their use of APIs has accelerated over the past 12 months, and that it will continue to accelerate during the next 12 months. APIs aren't really a new tech-nology; they've actually been around for a long time. What is new is that they are in far broader usage today than they have been in the past. As of late 2015, the *ProgrammableWeb* API directory listed more than 14,000 APIs.

When thinking about API strategy, a number of important questions arise. Why are APIs so well used and what effect might they have on your business? What is the impact of the expanding API economy on both providers and consumers? What are API tools and how can an investment in tools be an extremely cost efficient foundation for pro-duction grade API use cases?

The causes of the API explosion

An API is simply code that is pre-written to enable interoperability between diverse systems and software. When using APIs, no one has to understand how the system on the other side actually functions. APIs allow anyone to access those systems with only working knowledge of the API itself. This is in contrast to the integration technologies of the past, in which custom programs had to be created for each integration use case.

APIs offer a much faster and easier way to connect diverse systems. This is becoming increasingly important for companies which are, or want to become, software-driven businesses. consumer-facing applications, public cloud, and mobile applications are all driving growth in API usage, as are social media and the internet. APIs offer a way to share raw data and processed information from the cloud platform, with traditional back-end enterprise applications and databases. This sharing of data reflects the importance of the interoperability of systems to the software-driven business. Although the techniques behind API-led connectivity differ from those supporting traditional data integration and ESB, or enterprise service based platforms, all represent interoperability. The number one driver for interoperability is the need to interact with business partners, such as suppliers, providers, customers, or partners. APIs make this faster, easier, and cheaper.

The widespread use of APIs covers nearly every industry vertical because the need for speed and agility has become a universal enterprise requirement; APIs have become a mainstay supporting high velocity integrations. Before APIs, connecting a computer in the data center with a scanner, for example, required a custom project and weeks or months of effort. Essentially a programmer wrote custom software to enable the two systems, which spoke different protocol languages, to interoperate. As recently as seven or eight years ago, in the early days of component-based service-oriented architectures, each integration was still essentially a separate product. From our perspective today, that was a very slow way of doing business and one which today's companies really can't afford to tolerate. APIs offer a more standardized approach that makes brittle organizational borders more flexible.

The rapid changes we're seeing today across virtually every industry vertical are made possible to a great extent by the API economy. Companies can extend their borders to new geographies, suppliers, devices, or customers far faster than they could in the past. And, with the coming of internet-sustained and similar hyper-integrated ecosystems, it is clear that API usage will only continue to grow over time.

The impact on API consumers and providers

Most companies actually start out as API consumers. That is, they access APIs provided by other organizations. For example, SaaS or cloud infrastructure customers are familiar with the APIs available from their cloud service vendors. Some, but not all, companies are API providers as well. These companies create APIs that provide access to their applications or services. Provider-written APIs can benefit either internal or external customers. One common use case for internal customers, for example, is to develop common data access APIs for developers to use in creating new applications. Such APIs can reduce development time by inventing the wheel only once, as opposed to reinventing it for every new application.

There are a few things that consumers and providers of APIs need to be aware of. The first is basic API readiness. With no discipline or governance, API usage can easily spin out of control.
Developing policies, standards, and governance up front helps to ensure that APIs are created and utilized in the most cost-efficient way possible. API management typically refers to not only the governance aspects that apply to versioning APIs, but also the application of external aspects that are beyond the API's description and implementation: security, SLA tiers around the API, registration models around the API, the collection of both usage and operational matrix data which then feed back into improving the API, and the introduction of new versions.

We also recommend that the APIs you use are well-implemented. This means that you need to make sure that the API is built robustly. It must be built on a reliable, scalable, and highly available foundation and it must be easily integrated with your existing back-end APIs and services and assets, whether legacy assets or not, that you already have. All of this leads to the API-led connectivity concept.

The ideal state for APIs

Companies of a certain size already have a large number of APIs. APIs are multidimensional in that they touch lines of business and IT top level management. While it's relatively easy to provide or consume APIs, the number of API connections that an organization uses tends to grow very rapidly. As more users and more applications connect, and as development creates new APIs and versions, the API landscape can start to look more like a maze to be navigated than the straightforward way to flexibly extend the business.

As previously noted, IT is transforming into an enabler of technology within different business units; one of its key new roles is to promote existing APIs from a hidden shadow state into a published, ideal state where there is much more reliability, availability, and discoverability. Also, IT not only allows the move of existing APIs into the ideal state, but also enables the incubation of new APIs to start in the ideal state by merging existing best practices, existing characteristics and learnings within the organizations around the creation of APIs, and, importantly, reusing existing APIs instead of reinventing them.

An API in an ideal state has certain important properties. First and foremost, the API needs to be designed in such a way that the consumers of the API will actually want to interact with it. The better designed that API is, the more easily it makes the data behind that interface accessible to the consumers and therefore makes it successful more rapidly.

Consistency is another important characteristic. For example, often a particular security scheme has been implemented five different ways or sometimes many more within an enterprise. That makes the API consumers' lives very difficult, but it's difficult for API providers as well because then they have to go and reinvent the wheel every single time around implementing a security scheme. Another example might be a business need to create an application that consumes three different APIs; each of them has the same security scheme but it's defined in three different ways. Three different site variations of software have to be created, and therefore interaction with that API in order to implement the application quickly becomes unmanageable. Consistent design enabled by IT is key to make API-led connectivity work.

Discoverability for application developers or API consumers is also important. Developers need to be able to easily find what APIs within the organization are available for them to be successful. But in many cases, there are only various spreadsheets or various Wiki pages spread around documenting that there's an API. What companies really need is a central place where all the APIs are actually exposed and can easily be searched and used from there.

It's important to note that this should be complementary to the existing architecture. It's all about reusing existing assets and business logic. What API-led connectivity does is transform the present investment in order to make it more agile and efficient. You can build on top of your IT architecture, whether it's assets, mainframes, data and databases or big data stores, legacy middleware that might already exist, or any custom applications. These assets present very valuable sources of information within the organization. The challenge is to enable the organization, the business units, and your IT organization to leverage this data and present it in an easily consumable manner to the application developers and, ultimately, to the stakeholders who need access to this information in a context-dependent basis. These stakeholders could be your employees, your customers, a general audience, or your partners.

A layered approach to APIs

These principles ultimately will help you create an easily consumable interface to those application developers that create touchpoints for the different types of audiences and the assets of an organization. As discussed earlier, in order to function optimally, businesses need to organize their APIs into these three different layers.

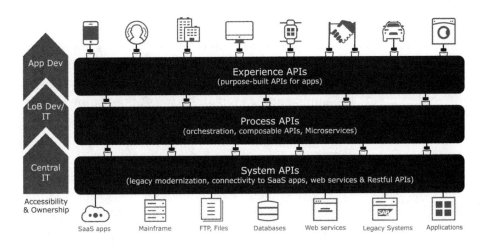

The system APIs are the APIs that expose the APIs of record. They expose the data information of the assets in a very easily consumable manner. Typically these APIs actually are the ones that change more slowly. At the top, the experience APIs are the ones that change at the fastest pace. That is because applications come to life at a much quicker rate than systems in the back-end. And experience APIs expose the data from the back-end data sources in exactly the right form for the application or devices in question that need to consume the data. In the middle are process APIs. This is the orchestration and composition that need to take place with system APIs in order to provide that data in a new way using business logic through the experience APIs.

If this layering is implemented using an API designed-first approach, that provides the ultimate agility and efficiency in order to make application developers to make API consumers successful and also to make the API providers extremely successful with the proper governance that allows this type of solution to scale.

It's crucial, when thinking about API-led connectivity, to first consider the design of the API. Whether that API is actually built on top of existing back-ends or whether it's built on top of existing APIs, it is important to start with success in mind. What does a successful interface look like? Once the outcome is imagined, then the API can be simulated.

This allows a rapid integration between API providers and API consumers in order to refine that interface definition to get it to the right point. This can be further validated through the automatic creation of portals and then, finally publishing of that API without any implementation so far, but for it to be actually usable so that the API consumers and the API providers can now go off in parallel to implement the API as well as create the application on top of the APIs at the same time in order to gain maximum efficiency.

Once the interface has been defined, the API developers can implement the APIs. In most cases the implementation of APIs actually doesn't come down to new business logic, but rather to the orchestration of existing business logic, or connectivity to existing business logic—in other words, integration. That mocked API can be realigned by making an implementation very easily within the uniform platform by having that interface lead to first a skeleton implementation as based off the interface. And then it can be implemented the different resource methods and wired up with the right orchestration in the middle with the different back ends.

Setting up API management and governance

The right tools can help provide visibility to all stakeholders while supporting the day-to-day operations of the API ecosystem. They can help to rein in any complexity and confusion. If you are an API consumer for example, things like infrastructure, application, and version level changes are important. One day you're using a company's APIs and everything is fine. The next day the integration suddenly stops working. This is not an uncommon occurrence.

Has something changed on your side? Is there a problem with your application, your network, or your data center? Or, has the API provider updated the API or the systems it connects to? And, without the right tools, answering these questions can take a very long time. It's also important to know who within your organization is using APIs. Who's accessing paid services and subscriptions and how much is it costing? Connections by unauthorized users can prove to be quite expensive. It's also important to clearly understand which users have access to the information provided by the APIs you consume.

If you're in healthcare, for example, you probably don't want to give every user in your organization access to patients' health records that come from external health service providers. Without the right tools in place, issues such as these can definitely harm the business, and tools help ensure that this doesn't happen. API providers have many of the same concerns, but they have additional issues as well. One is managing the API lifecycle. As your applications and data schemes change, for example, your APIs will need to change as well. This means you often have an API life cycle operating alongside your existing application development life cycle.

Both must be in sync to ensure that API changes reflect the latest changes in application code and data scheming. Access control becomes increasingly important particularly if APIs you provide are accessing production data, which is almost always the case. Security becomes a sensitive issue when anyone with a cell phone starts to access your mainframe. In addition, performance and availability management capabilities are an essential function for API providers, but this is often overlooked. Integrated applications have multiple moving parts and performance or availability problems with even one component can impact the entire delivery chain. This means that I'm a customer connecting to your internal application using your API, your performance problems suddenly become my performance problems as well. So while, in the past, down time may have cost your company so many dollars per hour, now your downtime is costing me time and money as well.

There are other advantages to good API management and governance as well. First, it ensures that API use complies with organizational and governmental standards because the compliance standards in a lot of industries are very strict. Also, good API governance and management can ensure that all versions stay in sync. It's also advisable to have a single point of control for all things API related. On this point, many companies today have API managers who can either be part of IT or lines of business, who oversee and coordinate multiple aspects of API utilization. Coordination of factors such as security, access and user management is important to mitigate the risks associated with providing or consuming the wrong data. Post planning is important as well.

API management tools

How can you choose the right tools to make sure your APIs are well managed and governed? First, look to find where your gaps are. Are you providing or consuming APIs, or are you doing both? Your tools needs will vary depending on which is the case for you. In addition to the security and user related issues, can you integrate data from your API focused tools with that of other enterprise data management solutions such as application performance and life cycle management products?

Application performance integrations are particularly important for providers for multiple reasons. Capacity planning becomes a concern as API usage accelerates as well. Wildly successful APIs present a good news/bad news scenario. It's nice to be successful, however successful

APIs also need higher rates of connectivity and higher levels of resource utilization on your end. As your company becomes increasingly reliant on APIs, it becomes increasingly important to make sure that you can adequately support some of these issues.

A single platform to manage API-led connectivity has a number of advantages. It's much more effective and efficient to manage and govern APIs if the capabilities—design, implementation, monitoring, operational monitoring, etc.—are accessible under a single platform. For example, imagine that suddenly there's an operational alert that shows that there are SQL errors going on within the database systems API. Well, that's probably a cause of alarm for many reasons, one of which could be that there is security threat. Therefore, the first thing to do work through the API management layer using the same identity and the same interface and the same organizational structure. You can apply a SQL injection policy and protect the interface as a first step. You can then go ahead to the portal for that API and notify all users that this might have happened, and notify the API owner that this might have happened.

Then you can actually start testing the fact that that policy has taken place through the API console within the portal. You can make sure that you're applying policy violation policies on the analytics side to make certain you're getting notified whenever that policy is violated so you can further investigate. This is just one use case that brings the implementation, the monitoring, and the consumption aspects of an API level together. If you had three different systems that you had to go to to address this scenario, it could be much more complicated.

APIs can provide a wide variety of business benefits. They can enable a company to become more efficient in delivering new products and services. They provide a simpler way to integrate compared to the custom integration and complicated metaware of the past. They enable businesses to become more agile and flexible in their interactions with customers, partners, and suppliers. They help support modernization, enabling companies to very quickly adapt to changing technology and business requirements. But like any powerful tool, you have to make sure you're ready to use APIs. Failure to plan for the processes, tools, and responsibilities which APIs entail can be a fatal mistake. Choosing the right tools can help maximize the benefits of APIs while minimizing potential risks.

CHAPTER 9

The rising value of APIs

Ross Mason, MuleSoft, and Joe McKendrick, Forbes staff writer

Data is, in many ways, one of the most valuable assets a business has. A growing number of consumers and businesses are incorporating web and mobile apps into their daily routines, and companies are using data to provide more personalized, tailored experiences to their customers. In addition, companies are analyzing customer and operational behaviour to make better decisions. These are some of the valuable new uses for previously isolated data sources.

APIs have emerged as as the most accessible way for consumers within the business to extract value out of that data; developers can use them to create new business opportunities; improve existing products, systems, and operations; and develop innovative business models. Analysts can grab new data sources more quickly and pull the data into their analytics platforms. As the keys to unlocking precious enterprise data, APIs need to be combined with enterprise connectivity to actually free the data from systems. The APIs is the piece that makes the data consumable and reusable, thus they become ever more valuable to business.

In 2015 MuleSoft surveyed 300 IT leaders about their use of technology, and the results were very clear—APIs and their integration possibilities are providing real value to the business, whether it's implementing cloud applications for increased agility to actually creating revenue streams.

We expect the value of APIs to the enterprise to increase as new ways are discovered to use data. Every industry and every customer touchpoint will find itself interacting with APIs, as developers further implement the orchestration and presentation of valuable data. APIs are transforming modern business, and we are starting to see companies capitalize on the opportunities that they provide.

How APIs are changing business

IoT

We anticipate seeing interesting Internet of Things use cases come to life, rather than major steps forward in devices themselves. Hundreds of new IoT devices will be released in the coming months, but it won't be the devices themselves that make waves. It will be the clever use of those devices—and their accompanying APIs—to generate value. For instance, 90-year old pest control firm Rentokil connects its mousetraps through IoT technology, and has increased operational efficiency through the automatic notifications of a caught animal and its size.

Overall, the key IoT theme will be identifying the value niches within industries that can benefit from this technology rather than trying to change the entire industry. For healthcare, it will likely be connected patients. For retail, it will be around making stronger connections between traditional and digital shopping. Behind all of these services APIs provide the link between the devices and digital services.

Cloud

When it comes to the cloud, enterprises are in an awkward tween stage — somewhere between the old world and new. We know that CIOs will continue to adopt cloud applications and seek better ways to connect on-premises systems and the cloud. Hybrid IT is now the reality for many enterprises and many are going through a refresh of their platforms, both business and technology. They are looking for scalable ways to connect and move data to the cloud, on-premises and back again as needed. There is a big emphasis on APIs to unlock data and capabilities in a reusable way, with many companies looking to run their APIs in the cloud and in the data center. On-premises APIs offer a seamless way to unlock legacy systems and connect them with cloud applications, which is crucial for businesses who want to make a cloud- first strategy a reality. More businesses will run their APIs in the cloud, providing elasticity to better cope with spikes in demand and make efficient connections, enabling them to adapt and innovate faster than competition.

In the MuleSoft connectivity benchmark report[6], we found IT leaders' biggest integration priority was cloud software and applications.

Omnichannel strategy

We anticipate many industries will turn to an omnichannel strategy to attract and retain customers by creating improved consumer experiences. By connecting the physical world with the online world, companies can bring new value and increase revenue opportunities. In particular, the retail industry will embrace an online-offline approach to increase sales. E-commerce stores will turn to a complementary brick and mortar store strategy, attempting to bring online shoppers in-store with exclusive offerings and deals, or add value by offering a unique experience beyond the ability to purchase in person. One example is eyewear retailer Warby Parker. It offers convenience and choice to its customers through a huge online selection, but it also provides custom fittings or repairs in their brick and mortar stores. Another industry that will take advantage of an omnichannel approach is financial services, which will look for ways to bring new products and services to market quicker through digital channels. This will mean improved mobile banking, faster payments and new consumer products. No matter the industry, companies turning to an omnichannel strategy will rely on APIs to create a link between cloud,on- premises systems and mobile, offering a seamless experience for their customers.

Changing role of the CIO

We will see CIOs shift from traditional IT delivery models to delivering capabilities to their business, allowing the consumers of these capa-

---

[6] MuleSoft Connectivity Benchmark Report, July 2015

bilities to build their own applications and processes. This is the decentralization of IT, where IT no longer owns the applications but are governors of the data. This will contribute to the expanding partnership between business and IT. CIOs are beginning to embrace their new role as a business enabler and are gaining confidence in doing things differently. They recognize their role is no longer just about keeping the lights on and the networks running. For this reason, successful CIOs will come to the table with a vision that helps put the company on a course of action toward greater digital transformation. The key step will be decentralizing IT by opening up APIs to developers and analysts, so they can gain access to reusable data. Additionally, IT will standardize on business and technology platforms to reduce their technology footprint.

Rise of the API economy

More enterprises are going to adopt an API strategy, with the goal of enabling greater agility and efficiency within their organizations and driving more innovation to compete with emerging startups that continue to erode their value proposition. This year companies, like Uber and Slack, achieved major success through their open API approach, and we'll see other businesses start to follow a similar strategy. First, traditional enterprises will open up APIs internally to break down information silos and unlock data. The next natural step will be for enterprises to open up those APIs to third parties, creating new revenue channels. For instance, in our global survey of IT leaders, 80 percent of large enterprises (10,000 employees or more) said that their company currently makes more than $5 million a year from APIs. In the same survey, the IT decision makers said an API strategy was one of the top three priorities to an organization's business plans in the next year. We're only going to see this increase in the coming year as organizations embrace the API economy and recognize its business value.

APIs are changing the equation

As more and more APIs come into use, the architecture underpinning them needs to evolve as well – organizations cannot simply attempt to deploy APIs on top of existing monolithic systems and processes and expect overnight transformation. Rather, the transformation begins with initiatives targeted at new innovative directions for the organization, such as the embrace of microservices, mobile apps, and laying the groundwork for a world of connected sensors. Above all, embracing APIs will help ensure that these connections are made intelligently and efficiently.

Our Connectivity Benchmark Report[7] confirms that 72% of enterprises have API strategies, indicating the importance of these building blocks for the composable enterprise. There's a direct connection to business value as well – generating revenue is considered the most important value that APIs provide to the business. More than 50% of respondents either are now generating revenue through APIs or will be within a year. In addition, 80% of large enterprises (10,000 employees or more) say that their companies already make more than $5 million a year from APIs.

What is the most important value that APIs add to the business?

Generate revenue 30%
Enable affiliates 21%
Engage external developers 19%
Enable partners 18%
Agility 7%
Drive innovation 3%
Enable applications 1%

What is your company's timeline for generating revenue through APIs?

Already generate revenue through APIs 19%
Within one year 54%
More than one year 20%
Don't know 8%

While revenue generation is an important part of the story, the impact of APIs goes much further into organizations, enabling transformation and agility at many levels. APIs enable enterprises to deploy apps

---

[7] MuleSoft Connectivity Benchmark Report, July 2015

quickly, in a repeatable way, which leads to a faster pace of delivery, and the ability to create new and innovative experiences quickly. In addition, APIs can greatly reduce the cost of change, enabling IT and application owners to change apps with minimal impact – especially when there are numerous back-end integrations involved. This is critical to agility since the pace of change of the front end applications is much faster than in the back-end applications. APIs also help enterprises achieve operational efficiency, enabling greater visibility and expanded capabilities since every API call from the mobile app to the back-end system is tracked and traced through an API key.

How companies add value with APIs

Tom Quinn, the CIO of News Corporation Australia, realizes the value of APIs to his business. "I think the future of the API journey is still yet to unfold – but I know for sure that it is the future," says Quinn. "I don't want to have a data center. I want it all gone. Currently, we have 60 percent of our computer in the cloud, but I would be happy if it were 100 percent. Our API platform is critical to this. It will allow us to chop and change funding and software when we know which we do know, as markets mature, so does the software mature. The future of technology for us is our API platform."

At a major financial analyst firm, which provides regulatory and tax data to professionals, there is a pressing need for better integration across its disparate systems, which was addressed through APIs. "We lacked a unified infrastructure, with disparate applications in a lot of silos," says this company's director of technology. "We needed to bridge these silos with a simplistic solution, to deliver actionable reports to our customers."

Organizations can free themselves from the limitations of their legacy systems so that they can start to change the way they deliver digital products and services to customers, partners and employees to fully engage in the digital economy. For example, New Zealand Post's parcel and courier business is run on legacy systems that track parcels, provide rates, and provide shipping options. "We realized we needed to abstract away from those underlying legacy systems, and provide interfaces that more modern and current developers could actually use," says Joe Brophy. "That led us pretty quickly to APIs and API technology."

At a US roof manufacturer, a concerted effort is underway to expand APIs to its partners and customers. The manufacturer built its API strategy on its extensive infrastructure of on-premises systems, which support everything from ERP to distribution to customer service. The goal was to keep these APIs as agile as possible, says the company's senior architect. "In manufacturing, IT is not the biggest part of IT organizations," he explains. "The IT organization has to be very lean. We needed a solution that is nimble and fast to support that." The manufacturing team's approach is to build and deploy APIs that are reusable, and can grow as the functions behind them grow in sophistication.

New Zealand Post Digital moved into the digital arena with three levels of APIs, targeted at enhancing its existing internal business, as well as branching out into new areas. "The first API is for e-commerce, logistics, and parcel delivery," says Joe Brophy, solutions development manager for New Zealand Post Digital. "The second one was to expose our addressing assets, for services such as data cleansing, credit card applications or for identity verification." The third, he continues, called Connect, serves to enable transfer and management of digital content.

Early on, Brophy and his team recognized "that APIs had a much more strategic role to play," bringing the organization closer to its customers, as well as development communities. The APIs have enabled the postal service's digital and IT teams "to stay close to the market and keep innovating. We launched the APIs to enable us to move at a faster rate, and make it easier to use our services."

For News Corp, APIs are bringing the company's products closer to its customers. "We're able to build separate, individual APIs for each of our smartphone apps to make the creation of the end products very, very easy," says Quinn. "The benefit to customers is that they get the content they need when they need it, on their chosen device. It speeds up the process. Without APIs, we'd have to run it one system, and that would require a lot of people doing a lot of work inside old, antiquated, slow moving systems. Now, we can atomize down our content creation delivery process, and our API platform helps us put it all together."

Unlock the power of APIs for your business

Businesses from every industry are using APIs to add additional value, from increased revenue to increased agility to improved customer experience. Extraordinary changes are taking place in the enterprise which necessitate a new organization and philosophy for utilizing technology.

CHAPTER 10

Developing your mobile strategy

by Priya Sony, MuleSoft

Mobile on the rise

There's no question that mobile applications are dominating both B2B and consumer experiences, and there's no denying the shift toward mobile taking place in the enterprise. Mobile is fast evolving into the primary channel for marketing promotions, customer community development, customer service, and support, supply chain management, manufacturing operations, and of course, digital commerce. Having an enterprise mobile strategy – once optional – has now become business critical.

We now review the primary drivers for mobile initiatives, the challenges faced by CIOs trying to initiate or support a mobile strategy, and the transformation stories of three market leaders who enabled mobility via API-led connectivity.

What are the primary drivers for mobile initiatives?

"The primary goal of mobile app initiatives is to either generate revenue (64%) or to improve the mobile experience of existing apps (58%)."[8]

---

[8] Survey Finds Mobile App Backlog Directly Affecting Enterprise Revenue. Rep. OutSystems, Oct. 2014.

Companies are building mobile applications with three main objectives:

- Enable employee productivity

  Businesses are building mobile applications to enable greater employee productivity. When employees can access important sales, customer, product, or operations data via mobile business applications, they spend more time working and making decisions on the go, and less time catching up in the office.

- Increase partner collaboration

  Businesses are also creating mobile applications to simplify and streamline interactions with partners and suppliers. Mobile applications offer instantaneous communication, making it easy for all stakeholders in the supply chain to stay looped into exactly how, where and when to turn their cog in the machine.

- Improve customer experiences

  Consumer apps have created an expectation for incredible mobile experiences, even for business apps, and businesses need to deliver it seamlessly and securely. Mobile customer apps increase brand preference by enabling customers to shop, compare, buy, and access services at any time.

CIOs are charged with enabling mobile initiatives across different lines of business in order to remain competitive and innovative. The need for mobile is urgent, but significant IT challenges stand between the CIO and a robust mobile strategy.

Mobility challenges faced by CIOs

Speed - Deliver the 100th app like the first

Mobile IT strategies don't just need to support four or five mobile applications—they have to work for a number of different business groups that are asking for multiple applications each. As each group changes strategies and systems, IT must also be able to quickly and seamlessly make updates to mobile applications. The lines of business can't wait six months to a year for an application to be developed or updated if they want to remain competitive—they are demanding multiple mobile applications for employees, partners, and customers at the speed of the business. Enterprises need to run fast on their first mobile initiative, and just as fast on the next 100 projects.

Deliver great applications, quickly

Success for a mobile IT strategy is very tightly tied to the speed at which mobile applications can be created and updated. Speedy mobile deployments come from two things—fast front-end development and fast back-end data access.

Front-end speed

On the front-end, the mobile application developer is focused on speed, design, and user experience. Mobile developers and architects are focused on deploying functional mobile applications with easy-to-use interfaces for immersive and responsive experiences. They are not necessarily aware that a lack of fast and secure access to data from various back-end sources in the enterprise will in fact hamper the project timeline and the ultimate robustness of Typically, the mobile application developer is focused on speed and design, and is either not aware of or doesn't fuss over back-end connectivity.

Back-end speed

The challenge of speed primarily lies in secure access to back-end data. Speed on the front end doesn't matter if an application's intended content is locked away in systems across the enterprise. Although a few mobile projects might be possible on an ad hoc basis, custom code and point-to-point integration not only slow app development, they create a brittle infrastructure and increase security risks. Architects and back-end API developers need to enable secure, self-served access to data from different enterprise systems to multiple mobile applications – in a scalable way.

Governance and control

Businesses need to deliver robust applications quickly, but they also need to ensure they do so securely. Exposing an enterprise's assets is risky business, and the greater the number of applications, users, and systems, the greater the risk of assets being compromised. Moreover, systems that aren't built to handle the volume of data requests as might be expected from mobile applications are prone to failure, down-time, and ultimately a bad user experience. To ensure a stable and reliable environment, APIs need to be designed with an understanding of the back-end systems and then built to deliver on those requirements.

This creates a conflict between the mobile application developer's need to access data quickly, and the back-end developer's need to ensure that access to enterprise data is well secured, governed, and managed.

The opportunity – API-led connectivity

The solution to solving these challenges is API-led connectivity. The fundamental building blocks of this architecture are purpose-driven development of APIs in order to meet application requirements, while establishing policies and managing access to back-end data.

API-led connectivity enables:

- Ubiquitous connectivity – Connect mobile applications to any source of data in the enterprise quickly and scalably.

- Fast deployments, fast changes – Self-service API access and composition enables developers to move fast, as often as they need.

- Scalable IT architecture – Expose back-end data to app developers by loosely coupling systems and without creating brittle point-to-point integrations

Why API-led connectivity enables mobile strategies

In order to support front-end speed while having robust back-end governance, enterprises need to provide mobile developers with self- service access to data across the organization. APIs help unlock data and assets by providing a layer of abstraction and control between mission-critical back-end systems and the front-end services being exposed to mobile developers.

APIs enable the speed and flexibility necessary to quickly expose all sorts of data to mobile applications. Composable APIs allow developers to quickly create new APIs from existing building blocks, ensuring fast access to everything in the enterprise. With APIs sitting between front- end applications and back-end systems, any changes made to the back- end won't affect connections to mobile applications.

API-led connectivity for mobile in action

Case study: enabling employee productivity

A large US food company was looking for a better way to enable their field sales teams. Field sales needed fast access to all their customers, inventory, and order information at their fingertips in order to sell better on the road. Moreover, the sales organization was spending far too much time on administrative and planning tasks.

The company turned to MuleSoft to help them take their mobile strategy to the next level. Working together, the large food company was able to provide a robust mobile solution for their sales teams that integrated information across 24 applications, including Salesforce and SAP.

With Anypoint Platform for Mobile, the food company's sales teams were able to:

- Spend more time at stores and decrease the amount of time spent on administrative tasks. Instead of being behind the desk, employees were out in the field selling. As a result, they were able to visit a greater number of stores, increase revenue opportunities, and reduce overtime hours.

- Make more informed decisions thanks to instant access to customer data, allowing for better sales conversations.

- Eliminate the need and cost for laptops in the field—data and processes were all made accessible through a robust mobile application.

- Accumulate 67,000 additional selling hours per year across 1,600 sales reps.

<u>Case study: increase partner collaboration</u>

A Fortune 500 beverage company came to MuleSoft with a backlog of mobile applications requests, strong interest in SaaS adoption, and a business strategy for growth through acquisitions. These requirements had put significant stress on the business' IT organization, which needed to improve speed and agility in supporting business initiatives, while still delivering on cost reduction targets. Caught in the middle of industry trends that were greatly impacting their revenue and market share, the beverage company looked for ways to remain innovative, better understand their consumers, expand their global and regional brands, and improve their operational efficiency.

Back-end Issues

- Point-to-point integration and custom code across the ecosystem and directly within the application, with numerous middleware tools sprinkled throughout
- A complex and brittle web of tightly coupled interdependent systems made making modifications difficult. Any changes required significant investments of time and resources.
- Numerous legacy integration technologies and the emergence of SaaS and mobile applications exacerbated all these problems.

Solution:

- The beverage company has launched an API-led connectivity rollout to increase agility in delivering mobile applications rapidly and with high frequency. They introduced their first mobile application in just three weeks—something that would have taken months previously.

- They set out to create a mobile application to automate and digitize wholesale ordering and streamline operations with partners. Through APIs created, managed, and monitored on MuleSoft's Anypoint Platform for APIs, the company was able to track all their external assets in the field, enabling them to know inventory stock situations and begin to address them on a timely basis.

- Additionally, central IT is now getting requests for awesome mobile applications from departments across the whole organization—marketing and supply chain operations, to name a few.

Case study: improving customer experiences

Over the past decade, a leading U.S convenience store chain – ranked as one of the largest private companies in the U.S. by Forbes – has been consistently praised for customer-centric innovation. Understanding the high demands of their customer base, the convenience store chain knew that a misstep in launching their first mobile application would negatively impact their growing brand.

As competitors like Starbucks and Dunkin Donuts started winning repeat business by tying customer loyalty programs to mobile applications, the convenience store identified an opportunity to innovate once again.

The company's President and CEO described their goals for mobile: "We want to integrate our app completely with the experience at the store level." Executing this vision, however, would require both business and IT to rethink their approach to connectivity. Prior to building the application, the team set out to identify how this application would enhance the in-store experience.

They landed on a few key capabilities. They wanted to allow customers to:

- Order and pay via the customer's mobile device
- Manage and redeem store rewards
- Check the prices of one product, gasoline, in real-time
- Check-in, view store hours, and get directions to nearby locations

To deliver a seamless customer experience, the mobile app needed to enable payment providers, loyalty vendors, and point-of-sale systems to all communicate securely with each other and with the store's back office systems and data. These communication points needed to be API-led and loosely coupled to provide flexibility should a vendor change. And with 81 disparate cloud and on-premises endpoints for application version 1.0 of the application, the chain's developers needed a solution that would make them hyper productive. Nothing they had in-house could connect these endpoints at the speed at which they were looking for.

By delivering this application with a vision for API-led connectivity on MuleSoft's Anypoint Platform, the convenience store chain now has the speed and agility it needs to accelerate its pace of innovation.

Delivering mobile experiences to enterprise audiences—employees, customers and partners—is an increasingly crucial business imperative, but has a number of technical challenges. Fortunately, the API-led connectivity approach to IT architecture can solve those challenges, providing businesses the agility they need.

CHAPTER 11

Expanding the frontiers of connectivity with the Internet of Things

by Ross Mason, MuleSoft

A company's success is now directly linked to how well it connects applications, data and devices. The way organizations compete today depends upon how efficiently they can do this. But the notion of what a "device" is has changed radically. Today, the term "devices" means anything connected that isn't a traditional web client. Sensors, connected machinery, street lighting, and appliances, among other things, are all now becoming connected devices. This shift to smart, connected devices is referred to as IoT, or the Internet of Things. We now wear devices, we use them for every daily activity, and they interact with our environment at home, at work and everywhere in between. These are becoming an increasingly important part of the fabric of everyday life.

The value that the enterprise gets out of these devices isn't contained within the devices themselves. Rather, their value to the enterprise is the data they collect and the way the enterprise either reacts to that data or uses it to create new services and products. We'll talk about some of our customers who are doing just that later in this chapter. Because the notion of connected devices is infiltrating many areas of our lives that we do see and many more that we don't, the conversations on IoT come from different angles, different spaces, and solutions for different things. There is a large spectrum of connected items ranging from the consumer through to enterprise to industrial use (the really big things).

	Consumer	Enterprise	Industrial
Devices	Wearables, home devices, stuff you can back on Kickstarter and Indiegogo	Commercial machines like cars, medical devices, billboards, vending machines	"Internet of Things that spin" – jet engines, oil pumps, turbines, industrial site mgmt
Value model	Value is in the device itself. The value to the consumer is in the Insights it provides (e.g. your fitness) and the experiences it enables (e.g comfortable temperatures at all times at home)	Value is in the digital services that device enables for better customer experiences such as loyalty programs or safety alerts for medical devices or for management/ maintenance such as equipment failures and auto refill order from vending machines	Value is in the instrumentation and analytics to reduce costs. In the industrial space, the cost of unscheduled downtime and manual maintenance checks runs into the hundreds of millions of dollars per year
Business model	Creating desirable devices purchased by consumers	Subscription service model (i.e. preemptive maintenance) or consumer engagement (i.e loyalty programs)	Platform solutions for running industry verticals i.e. factory management, pipeline monitoring

Enterprise IoT

IoT in the enterprise today comes down to using connected devices to drive outcomes in two main areas:

- Creating better (sometimes new) customer experiences
- Optimizing operational efficiency

To take a closer look at customer engagement, there are three types of models:

- Extend an existing business model—e.g. Amazon Dash, a smart button that allows you to order a refill on demand, taking the shopping basket into the home.
- Create a new business model—e.g. Briggs and Stratton's now offer service monitoring for devices like power generators that can alert you if they are low on fuel or need repair. This creates a new revenue stream and customer engagement model.
- Brand new products and services—e.g. Coca-Cola Freestyle machines offer consumers 50 flavors that they can mix to their own taste. This new consumer experience also provides important market research data that will help define new flavors.

For operational efficiency the use case models fall into a couple of categories:

- Instrumentation of machines and devices with sensors, making them connected devices that create data that can be analyzed for changes in state—e.g. monitoring a farming irrigation system that is managed based on weather patterns, rainfall, or season for a type of crop and air temperature.
- Efficient collection of data that already exists at remote locations— e.g. factory machinery created in the last 10 years create some amount of data but collecting that data and sending it to a central location for processing and analysis is difficult (and often manual) due to network and security limitations at remote locations.

In all these scenarios, the challenge of adopting an IoT strategy is partly about collecting the data from these devices and processing it, and partly about making that resulting data actionable and linking it to the operational systems that help you know your customer or your business and then putting that knowledge to use to drive a better customer experience or operational efficiency. So in a technical sense, IoT is a combination of processing big data in real-time and connecting that data to other things to make it actionable. Many organizations are not set up to do either of these well at scale, and the IoT technology domain creates some additional challenges around connectivity, reliability, and security that warrant an evolved approach for dealing with IoT architecture.

IoT represents an architectural shift

IoT is changing the computer hardware model that we've had for the past 30 or 40 years. Consider all the different phases of hardware models that have existed: green screen to mini-computer to PC and today to cloud and mobile. The computer architecture has been consistent—Client and Server. What's changed over the years are the clients. For IoT, however, there's a difference. There is actually a third hardware layer which breaks the traditional client-server model. Developers and architects are used to building software systems across two physical tiers. But for non-traditional web clients, there's a new notion of a third tier. This is an emerging concept, spearheaded by research at Princeton University, and it's called the Fog Layer or the Edge Layer. It is responsible for being the first line of connectivity for these devices to connect to before they go to the back-end systems — the server.

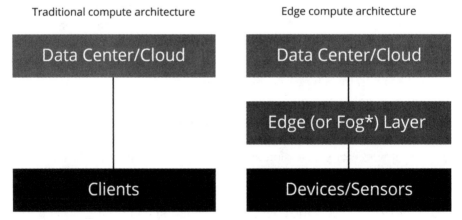

It has been predicted that there will be thirty to fifty billion connected devices in existence by 2020; as of 2015, there are already tens of billions of connected devices. If you look at what's on the roadmap for IoT, it becomes pretty clear that there's a very distinct role for this Edge Layer.

Let's demonstrate the Edge Layer by using the example of a connected building. In a such a building, the light bulbs and the air conditioning, climate control, blinds and other infrastructure are all running through connected devices. Let's say the building has fifty thousand connected light bulbs. Each bulb pings its status information to a local hub every 10 seconds plus any state change. That is a lot of repetitive data being generated. That amount of data exchange is not very efficient to send directly to the cloud, particularly if the building owner is paying for bandwidth. The role of the Edge Layer is to sit close to the physical location of the sensors and the devices and collect information from those devices, then collate that into more valuable data sets.

The new Edge Layer

The Edge Layer is responsible for connecting devices locally, and manages the data collection and connection to the server. The benefits of this approach are:

- **Data filtration:** First pass data filtering reduces the amount of data transmitted but retains the meaning of the data.
- **Connectivity protection:** Device connectivity doesn't fail if the network fails or there is an intermittent connection. The Edge Layer is responsible for handling outages and store and forward of data.
- **Site level management:** Enables site level orchestration across devices from different vendors using different protocols.
- **Device agnostic control:** Site abstraction allowing server/cloud application to be agnostic to the device implementation it controls.

The Edge Layer has three main components in a typical IoT deployment.

1. The device or sensor itself. In IoT this is the client that generates data and/or receives commands to execute.
2. Most devices will connect to a gateway that enables access to the internet or a private network. Typically these gateways

speak a proprietary protocol between the connected devices and then allow connectivity through the gateway using a standard protocol such as HTTP.

3. The Edge Controller is responsible for connecting to all the gateways and independent devices in a physical location. The Edge control collects and collates data from all the devices, transmits data, and accepts commands from the server to execute across some or all the devices.

Fig 1. MuleSoft's Reference Architecture for IoT applications covering the Client Layer, Edge Layer and Server Layer.

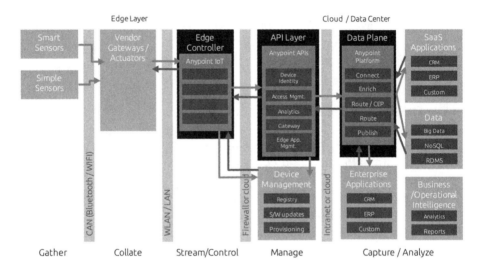

The Server layer is similar to what we already see in other client server architectures such as mobile. Many of the boxes in the Server Layer will seem familiar, because most of these components may already exist in your enterprise. The key pieces are:

- The API Layer is leveraged by IoT architecture to connect to the server layer. This provides consistency, control, governance, security and an analytics-based model for connecting external sites.
- Device management is responsible for knowing what devices are within the IoT network and sometimes is part of the authentication chain. Device management is also responsible for upgrading software on the Edge controller and possibly the gateways too.

- The data plane provides the event streaming, transformation orchestration and connectivity to the applications and systems that at can use the data coming from the Edge Layer.
- SaaS applications, Enterprise applications, Big Data and BI are typically the consumers of the data coming from the Edge Layer; they make it actionable through analytics, dashboards, and application processing.

Going back to the building management system, the light bulbs (devices) would connect to a vendor-specific gateway. That gateway is provided by a vendor and is the interface that allows other things to connect to it or provides the ability to get information to and from the devices. Typically the gateway uses a proprietary protocol to encode and communicate data from the the devices attached. The Edge Layer enables the collection of data from multiple different gateways and/or devices from different manufacturers. The building management system might have light bulbs from three or four different manufacturers in the building, so there will be three or four different types of gateways that need to connect. In addition, if the building management system wants to connect other devices from other systems like HVAC or environment controls, there's no easy way to have control events executing across those different types of devices. To turn all the lights off, a different command would need to be sent to each of the vendor gateways. The Edge Layer serves and a consistent control point for connecting to the gateways and translating commands for each of the vendors.

The Edge Layer is critically important for network reliability. The IoT reference architecture above is separated by network boundaries, so at the device side it could be Bluetooth, Zigbee, ZWave or one of many proprietary protocols between the device and gateway. The Edge Controller translates between the gateways to more standard or common protocols. Between the Edge Layer and and the Server Layer there is usually a private network IP network, 3G or public internet. Typically in most IoT applications, security constraints are actually defined by these network boundaries to control access and monitor.

What happens when network connectivity is lost to the Server Layer? The Edge controller serves as the point of contact for devices and gateways so that they can keep operating without connectivity to the Server

Layer over the network. For example, you are monitoring a piece of machinery watching for temperature fluctuation. If the temperature goes outside the prescribed boundaries there needs to be a series of events that triggers to alert other systems and to trigger an actuator to inject some coolant. The logic that decides whether the machine needs to be cooled is local to the machine itself so even if the network is down the machine will continue to function properly. If that logic was only accessible via the Server Layer and the network was down the machinery may overheat and break.

Another factor to consider is that the location of these devices and sensors can be extremely varied. Where connectivity is intermittent, perhaps in rural areas or countries where the availability of connectivity can be limited, the Edge Layer is crucial to build in redundancy while connectivity to the Server Layer is lost.

At its heart, IoT is about capturing and leveraging data being generated by connected devices; they create the physical Web, blurring the lines between how we interact between the physical and digital. The future for IoT seems promising with many IoT networks already connecting elements of the physical world. We already are able to monitor and report on traffic flows and warn of potential failures in machinery allowing pre-emptive maintenance to be conducted. Computers and personal connected devices in healthcare settings will constantly monitor patients and report any changes in their condition directly to their doctor. In the architecture I just described, the Edge Layer of the network will be responsible for performing collation of data and real-time event processing to allow automated tasks. This IoT architecture reduces the amount of data sent to back-end systems and provides a control interface that can access and manage local devices and sensors. The role of APIs at this Edge Layer is critical to providing easy access to connected devices, either through a hub or directly.

IoT in the enterprise today

The current trends that we're seeing in the enterprise suggest that advances in IoT won't be about the devices themselves, but rather the clever use of those devices to generate value. Companies getting the most success out of connected devices are identifying the value niches that can benefit from IoT technology. Let's look at some of the current examples of companies getting IoT value in the two enterprise areas we identified earlier, Creating better customer experiences and operating efficiency new customer experiences.

Creating new consumer experiences: Connecting with your pets

There are many IoT scenarios that do not require an Edge Layer, but can leverage the Server Layer architecture to connect devices. There's a group of people that loves their pet more than anything. i4c is a company that has created a pet collar to help those people be more connected to their pets with a collar that behaves like a FitBit for dogs. But creating this collar created a number of interesting problems. Unlike FitBit users, dogs don't have smartphones, so the collar can't connect to the internet through the phone. i4C had to come up with a way of connecting their collar directly to the cloud to send the data it was collecting. The collar was designed to talk directly to the internet over 3G and they used an API layer built on our cloud platform to create the data gateway that allowed the collar to connect with their back-end data collection and operational systems. At the time of writing they were approaching about 75,000 of these collars in production, and are collecting 5 billion data points weekly. This is where reliability of data interchange becomes important; funneling all that data and processing is only going to become a greater challenge as more and more devices add a greater amount to that load. Any platform that helps run IoT devices is going to need the infrastructure to scale the amount of data processed very quickly.

Operational efficiency: Benchmarking manufacturing

One of the largest CPG companies on the planet has a seemingly simple problem. They have about two hundred factories, and each one has a set of machines that gathers all sorts of information about the manufacturing process, how much material they are using, failure rates, etc. This company had no easy way of getting that data from those factory locations to their head office where they can use the data for real time reporting and ultimately benchmarking productivity between their factories. They would actually have to have people go and collect the data on a weekly cycle in order to understand what was happening in their factories.

Now, using Anypoint Edge™, this business has a cost-effective way to connect two hundred sites. Collecting that data through an Edge Layer allows them to get that information in real-time and allows them to get insights into what's going on on their manufacturing floors and processes instantly.

Operational efficiency: Reducing workload by 70%

Another company that uses API-led connectivity to power its IoT initiatives is Rentokil. If i4C is for animals you love, Rentokil is for animals that you don't love so much. It is a multinational hygiene services company. One of its services is pest control; they have about a thousand employees that check pest traps for rats and other undesirable animals. In the past, somebody would go check the traps every couple of days; but then humane pest traps were introduced, which meant the trap had to be checked 24 hours as an animal can't be left in there for longer than that. Therefore, the cost of running these traps is pretty high. So to bring down the cost and make their business more efficient, Rentokil created smart connected traps. They used a sensor on the trap that will send out a signal when the trap has caught an animal; then,the pest controller will simply go to empty the trap. This has increased the efficiency of that workforce by 70% since they no longer have to check empty traps and can optimise the routes taken by the pest controllers.

Conclusion

The ability to support devices beyond mobile applications within the enterprise is becoming increasingly important, with some industries running ahead, many in the planning phase. The reality for enterprise IoT is that it will extend your existing value propositions and help drive new products and experiences; devices will become woven into the fabric of your business. And it needs to connect with operational systems to link the data to CRM, ERP, or BI systems. Having the composable building blocks to enable information flows from new sources is a necessity. Being able to manage and govern those flows is important; for these systems to work, being able to define clean interfaces for data ingestion and then throttle access is mission critical to keep your old and new systems running together. You need composable building blocks to make this work. APIs are at the center of IoT architecture and need to be at the center of your IT strategy. The emergence of IoT has extended the notion that APIs can connect anything to everything. We believe that the value of IoT in the near future is the clever use of those devices—and their accompanying APIs—to generate real value for enterprise workforces and their customers. The innovation in IoT will come in industries identifying value niches where connected devices can benefit from this technology. The CIO's role in developing an IoT strategy comes in imagining the successful outcome and building an IoT strategy, using an API-led approach to enable it.

CHAPTER 12

Trends in API security

by David Berlind, Editor-in-Chief, *ProgrammableWeb*
and Aaron Landgraf, MuleSoft

Another way in which the role of the CIO and IT is shifting is providing enterprise IT security. Previously, IT was charted with connecting one back-end system to another and making sure those connections were secure. The security challenge and opportunity comes from connecting channels and assets to audiences, customers, partners or employees. There is a mandate in many businesses to create better experiences, delivered through applications, be they mobile apps, desktop apps that help employees collaborate, or microapps that accelerate partner onboarding.

The number of apps that IT has asked us to create has exploded, as have internal and external APIs. Ensuring these APIs are secure is often an afterthought. As Tim Prendergast, former Adobe Security Team Lead says, "there are far too many APIs being cranked out in such a short period of time... there is no way that they have all been properly secured and built. There will definitely be new attack vectors in an API-centric Internet, but we are still too early to know the pervasiveness of such attacks."[9]

In 2015, there were 208 million records exposed via API breaches, which cost companies over $32 million dollars. And customers are starting to pay attention. In a recent survey, 38% of customers said they would no longer do business with companies that suffered a data breach, and 46%

[9] Falco, Anthony." Voices: Tim Prendergast, Evident.io co-founder, on cloud security, API vulnerabilities." Orchestrate blog, 10/29/2013.

said they had advised friends and family to be careful sharing data with an affected organization, expanding these attacks' footprint.[10]

A particularly stark example of how dramatic the effects of these attacks can be is an incident in 2013 where the Associated Press' Twitter account was hacked and false tweets about the White House were posted, causing the Dow Jones to fall more than 140 points.
The Dow fell because algorithms monitoring national news patterns adjusted their portfolios based on these false tweets - all of which are enabled by APIs.

The effect of these breaches is quite stark and have consequences that IT professionals want to avoid. It is possible to take actionable steps to prevent API breaches, and the first is to understand the common patterns these breaches have.

Hacking "Micro-patterns"

When considering existing real-world security breaches, there are some commonalities about them that I call micro-patterns. These micro-patterns then add up to a larger pattern that often repeats itself.

- One example of a micro-pattern is the search for scale. In order for them to really be effective with what they do, whether they are trying to break into financial information or if they are trying to wreak havoc with something, scale is always something they are after. APIs are perfect for this because they are all about scale; the reason APIs are created in the first place is often to achieve some degree of scale in terms of interfacing with consistency. Therefore, they are sitting ducks for exploitation. APIs seem like they are perfectly built for hackers to break into.
- Another is the kind of system that is hacked. The original transgression will often be very targeted and unattended. Hackers will somehow find a way to penetrate the system to go exactly after a specific vulnerability, which often will go undetected for some time. If you follow any of these transgressions in mainstream

---

[10] Economist Intelligence Unit. Privacy Uncovered: Can Private Life Exist in the Digital Age? April 2013.

media you will notice that many times, the organization reporting the problem will say, "this has gone on for this period of time without us knowing that it was taking place."

- This creates another problem, which is common to these attacks. When companies don't know how long their systems have been compromised for, it's hard to know how much damage was done. Because these attacks very often leverage customer relationships, many people get pulled in through the social network in a way that spreads the effect of the attack. Hackers rely on trusting relationships that people have with their friend, family and business contacts on the various social networks to spread the vulnerability.

- These attacks can result in the publication, sale or even blackmail of some content. Hackers will also, in some cases, actually publish the source code. They are very proud of their work and often want to share it with others; if they had to write some code in order to achieve the hack they'll publish it and hope other hackers will try it on other sites, other organizations, or even try again with the original organization that got hacked.

- Another micro-pattern is that there is always media coverage of these hacks and, sadly, a lot of times people receive a lot of useless expert advice. For example, an influential media organization might cover a breach that becomes big news and suddenly there are a bunch of people who are chiming in as experts telling you how to solve the problem or what to do about it. Very often those opinions are misinformed.

- Eventually, depending on the size of the hack, some sort of company disclosure is forced about what went wrong and how many people might be affected. Everybody is expecting company disclosure; in some cases it's lacking, and in some cases companies don't even acknowledge that it's happened. You can see a variety of patterns on this one point. The news will often go viral on social media, with an attendant backlash from people telling their friends and family to watch out for a particular company or brand. Social media is not a good place for negative information about your brand to be spread as news spreads quickly and not always accurately.

- The final micro-pattern is that these attacks often open the door to security vulnerabilities at other companies. Once news gets out that someone has been hacked, attackers may also send phishing emails saying, "Hey, you may have been hacked by this, so you better change your password." These are not authentic messages from the organization that got attacked, but they increase the scale and reach of the initial breach.

An example attack - the celebrity iPhoto hack

The first example we're going to go over is what I call the celebrity iPhoto hack. This is when some very compromising and personal photos of various celebrities like Jennifer Lawrence, Kirsten Dunst, and professional baseball player Justin Verlander were released across the internet. The hackers leveraged an API that was used by Apple's Find My iPhone application. One of the most important pieces of this hack is that, as anybody who uses Apple products knows, there's one single sign-on for the entire Apple kingdom. You use that for iTunes, you use that for your iPhone, and you use that for your iPad.

That by itself created something of a vulnerability because once hackers get those particular credentials they have access to a lot of information. The Find my iPhone application relies on a private API. It is undocumented, but, generally speaking, the APIs that mobile applications use are relatively easy to interfere with and reverse engineer and thus can be a bit more easily compromised.

The API behind the Find my iPhone application only relied on the single sign-on credentials to get in. In this case the issue according to the hackers themselves was there was no rate limiting on that particular API. Rate limiting puts a cap on the number of times, for example, that you are allowed to try to enter a correct password. You might try to log into a service when you forget your password; you might try it once and it says no, that's the wrong password. You might try it again and it says, wait a minute, that's still the wrong password. The third time it starts increasing the challenges you have to overcome trying to log in. Now it starts to think you're maybe not the person you say you are. Eventually, it might lock you out of your account.

In this case this particular API had no rate limiting on it, which meant that it allowed the hackers to repeat an infinite number of user ID and password combinations until they got a hit. This is where hackers usually accomplish their objective of achieving scale because they only have to write a little bit of code to try something over and over and over and over again when there's no rate limiting in place. Sooner or later they are going to get a hit.

This is called a brute force attack where you try lots of user IDs and passwords. The hackers who developed the attack on Apple called it iGroup and they published their code to GitHub. This accomplished one of the micropatterns; when they published the code there was a chance that other hackers would go out and do whatever they wanted with it.

A lot of hacks involve the compromise of user IDs and passwords; unfortunately out on the internet there are huge databases of user IDs and passwords sitting there that hackers can use as the source information for credentials to try when they are doing a brute force attack. These hackers used the infamous RockYou database compromise, which has over 30 million accounts. They don't try any random user IDs and passwords; they go into a database like that and figure out the most common ones. They start with those and work their way through it until they get some hits, and invariably they do.

Once they got a few hits they used a piece of software called Elcomsoft Phone Password Breaker, which allows the attacker to actually get into the iPhone account or the Apple account and download all the content from it. The hackers had access not only to photos, but also text messages, calendars, address books, and phone call logs. This is a commercial piece of software that's available from Elcomsoft.

Then, within hours of the photos getting distributed across the internet, Apple put rate limiting on that API. It was too late to avoid the publicity, though. The news was spread all over the media, and this was when the phishing attacks happened. People received a message that looked very much like it came from Apple, which said, "Your account might have been compromised; you better go here and change your password." Of course, those phishing attacks were designed just to feed the hackers more legitimate user IDs and passwords.

Apple claimed that this was not a breach, which was surprising to many observers. Their official statement that this was a result of targeted hacks and phishing attempts. But according to the timeline of the attack, publicized across the internet and in many of the security forums, that was actually not the case. This was very much a breakdown in API security because there was no rate limiting.

They also advised all users to activate two-factor authentication so that a second factor, aside from just a password, was needed to log into an Apple account. Unfortunately, with API-based interaction, two-factor authentication may not always come into play. This is very important. You have to be careful about what you tell your customers and your users because with API-based interactions there's often not an opportunity for a user to provide a second factor. Two-factor authentication raises the barrier to entry in a great many cases, and everyone should be encouraged to enable two-factor authentication wherever it is available to them. However, it is by no means a silver bullet.

To recap, the micro-patterns this attack displayed:

- There was scale because the hackers were able to run their code virtually indiscriminately without any problems because of the lack of rate limiting.
- It was highly targeted and undetected; they knew exactly which API they were going after. They had discovered this vulnerability, and they were able to take advantage of the fact it could spend a great deal of time undetected.
- It resulted in the publication of content, so there were many compromising images distributed across the internet, and received significant media coverage, a lot of useless expert advice on what to do about it. A lot of people were out there saying, "Oh, you better protect yourself." A lot of the advice that was being given was incorrect.
- Apple was pressured to disclose, but didn't exactly, in my estimation, come forward with all of the details that would have allowed everybody to have a good understanding of what took place.
- Then the breach went viral on social media. The source code was published. There were subsequent phishing attacks and additional transgressions and also additional publications.

How Apple could have prevented the hack

What could have been done to stop this? The most obvious course of action was instituting rate limiting. That is apparent because Apple responded by installing rate limiting on that particular API, which basically stopped the hack dead in its tracks.

Apple could have also instituted secured communications between mobile apps and server hosting, and servers that are hosting underlying data. One of the challenges is that most communications between mobile applications and the servers that they talk to are easily exposed and reverse engineered, especially when that communication has taken place across a wi-fi network. One should always proceed in this area on the basis that those communications are vulnerable to reverse engineering and sweeping by hackers.

How could those channels be better secured? There are a number of techniques that you can do, but at the same time this is a very evolving area of API security where all the answers are not clear yet, but some things are being done to address the known weaknesses.

Let me issue more prescriptive guidance on credentialing. As previously mentioned, hackers leveraged the existing corpus of information in order to figure out what user IDs and passwords to try. This speaks to the fact that users are still widely allowed to pick very weak passwords. Everyone has been confronted by a password dialogue box or a password error message that your password is too short or lacks certain types of characters, but another area that needs to be explored is how can users reference other databases that have password data already in them and reject a potential password that has already been compromised.

This might be inconvenient for the user, but these recommendations are designed to minimize the risk to the service provider because these risks could have major consequences, not only financially, but also in terms of customer loyalty and brand application.

The Buffer attack

In October of 2013 Buffer, a social media service, was compromised. Buffer allows you to post one status update to multiple social media platforms like Twitter, Facebook, and LinkedIn. You write your update once, hit a button, and it is automatically distributed to multiple services. The attack on Buffer started by the breaching of a database belonging to Adobe, which revealed 150 million user IDs and also the hints to the passwords.

For example, a hint might be "last name of Grateful Dead artist" and that gives the hacker enough clues that it might be Jerry Garcia. In this way, they obtained enough information to get lots of user IDs and passwords; they started using that information assuming that people were using the same IDs and passwords for other services in order to execute their attack.

The hackers reverse engineered these passwords and then they targeted the developers that were working for Buffer. Eventually they were able to gain access to Buffer's private code repository on GitHub. From there they were able to explore the first code and find all sorts of keys and credentials to actually log into Twitter and Facebook and make posts on behalf of all Buffer's users.

Not only did they get the necessary information for a hacker to write an application and have the application pose as though it was Buffer, but they also targeted Buffer's database service provider, MongoHQ. They were able to gain access to MongoHQ's infrastructure where those databases were kept.

With this information, they were able to find the information that Buffer was storing for all of its users. Every time a new user comes into Buffer and says it's okay for you to publish something to my Twitter account on my behalf, Twitter will issue tip-offs and OAuths that are specific to that user. As result of this hack the attackers were able to get access to all of the OAuth tokens belonging to Buffer's users. Now the attackers can also not only act as Buffer's application, but they can also post to Buffer's user's Twitter accounts on those user's behalf, unbeknownst to the users.

Some of this data was stored in the support screen that MongoHQ staff had access to so the hackers likely wrote their own code to scrape the data off those screens. When they wrote the code to pose as Buffer, tens of thousands of Twitter and Facebook accounts were hit with a weight loss scam.

This is where the leveraging of trusted relationships can increase the scale and severity of an attack. If you, for example, were a friend of mine on Facebook and I suddenly posted something about how easy it was to lose weight, you might click on that link because you trusted me.

There is a high probability there was malware behind this link, designed to infect computer systems and then get access to other data. However, by the time it was all discovered the hackers had covered their tracks, so it wasn't clear what their motivations were. On October 26th, almost immediately after this, Buffer disclosed publicly exactly what happened. There was an ongoing investigation so as they learned more they published more. MongoHQ also disclosed, but not quite as much.

As a result of all of this and other transgressions Adobe finally sent out a message (after almost two months) telling users to reset the password for their accounts. I was talking to the CTO of Buffer and he shared some interesting information. In their investigation they had common IP addresses that were showing up across GitHub and MongoHQ. They could also see that the hackers were using their own Twitter accounts to publish this scam so they were able to spot which Twitter accounts were technically responsible for this.

Once Buffer turned on two-factor authentication for their code repository so that people couldn't get into the source code anymore, the hackers were asking on Twitter whether anybody knew how to get around two-factor authentication. Surprisingly, there was other information in the source code that the hackers didn't take advantage of— for example their Stripe (a credit card processing API) credentials.

Then we saw a near-identical repeat attack on Pinterest, although Pinterest was very tight-lipped about anything that had to do with multiple attacks on their infrastructure. But it was quite clear; the patterns, and the thumbprint of the attack were almost identical. The same weight loss scam was spread across Pinterest.

Micro-patterns of the Buffer attack

What were the patterns?

- The attackers achieved scale. They were able to publish these links to tens of thousands of Facebook and Twitter accounts.
- The transgression itself was highly targeted. They were going after very specific developers and infrastructures to get at the data that they needed.
- They leveraged trusted relationships.
- MongoHQ, Twitter, and Facebook all got sucked into the news. There were official company disclosures and the breach went viral on social media.

How Buffer stopped the attack

Buffer first instituted multi-factor authentication on the code repositories. One of the most common ways that hackers find their way to API vulnerabilities is to have Google alerts set up to continually call GitHub looking for people to publish their source code and the mistake ordinary people make with their API credentials.

When OAuth credentials are being stored in a database, make sure they are encrypted. Do not have them in clear text either. That was a mistake in this case. Don't sort credentials in the source code in plain text. Encourage developers' sensitivity to phishing attacks. Part of this web phishing attack was targeted to certain developers that were working for the various companies that were involved. They need to be continuously made aware of how important it is for them to be vigilant about preventing phishing attacks against their credentials. For example, they should never use a user ID and password on a source code repository that they are using on other servers.

Security for mobile APIs

There are a lot of different places where hackers can attack when a mobile application is talking to its back end. You can make an attack inside the device itself. This just shows the workflow, so that little gray box in the top left-hand corner is the application. It is usually talking through a HTTP because it's using a web API. The attacks can take place inside the device, it can take place on the Wi-Fi network or somewhere else in the cloud, or it can take place somewhere on the service provider's infrastructure. There are all sorts of credentials that can be discovered. Application secrets, OAuth keys, user IDs and passwords, and so one thing that's very important for all API providers to know is that a lot of this information is discoverable when a mobile application is talking to a back-end service.

There is an application called SSL that is usually downloadable from the Google Play store on an Android device. It will not only snoop in on the conversations taking place between a mobile application and the back-end service, but even if you attempt to encrypt all of that traffic using HTTPS it will unencrypt it for you. In the security world, this is called the man-in-the-middle attack; you trick the application into thinking that it's talking to the back-end service when it actually is not. The result of that is the application itself is able to encrypt everything that's going back and forth between the application and the back-end servers.

I actually ran this application and I was able to get user IDs and keys and reverse engineer how the application belonging to a very well-known brand worked. This very popular brand had an app talking to its back-end service. I essentially copied the data that was being passed back and forth, so that I could emulate the API. I changed some things to anonymize myself, for example, the latitude and longitude. Then I just used the code command from my computer to see if I could interact with the API as though I was the application itself. Sure enough, I was able to create all kinds of anonymous accounts on the service's infrastructure.

Now, when I did this I only created a few accounts. You can imagine that once I'm able to do this I could have written pieces of code to do this at scale. I could have completely brought their systems down by creating tens of thousands if not millions of accounts just ad hoc and consume all their storage and their resources. This is a good example of where again there was no rate limiting in place.

What's going on between the mobile application and the back-end service is very much unresolved at this point, though there is work being done to solve it. The application I ran may have been in my device and allowed me to spy on one particular application and only my credentials to that application; however, there's another version of this called MITM proxy (MITM stands for man in the middle), and you can run that on a public Wi-Fi network in a way that the people who attached to that would think they are attaching to some official Wi-Fi network when they are attaching to one that belongs to you. Then you can do the exact same thing that SSL does, you can encrypt the information going back and forth.

For example, at a conference with 5,000 people attending, somebody can hack in, create a hotspot that looks like the official conference hotspot, and decrypt all the information that's going back and forth between all of the users of that hotspot and all of the services that they are interacting with.

One solution to this is what's called certificate pinning. This is a stop-gap solution that specifies only when this application is talking to a back-end service it has to be to a specific certificate from a specific IP address. This breaks the idea of DNS, so it's a stop-gap. For ideal security, what you really want is DNS to be always working, because if you changed the location of your endpoints to a different IP address, in the future, the apps that use certificate pinning will work fine.

There is ongoing work at the IETF, the Internet Engineering Task Force, that will better secure the whole OAuth workflow. The service on the back end knows that the application that's saying, it has these credentials to talk to you, actually deserves to have those credentials and should be the one that's talking to you as opposed to the Apple attack where the API just thought the right application was talking to it and in reality it wasn't.

There is work being done to secure the OAuth workflow; however, it's not completely finished and it certainly hasn't found its way into the mainstream. The best practice is to become aware of the work and be sensitized to the necessary steps to best secure APIs. This is brand new work and it takes a while for it to find its way to the infrastructure.

The challenges of API security

There are a number of challenges that make securing your APIs difficult:

- There is a massive proliferation of APIs taking place where security is either an afterthought or a non-thought. There is a bunch of absurdity going on in the area of user IDs and passwords. There are too many weak and shared passwords out there.
- There are non-uniform implementations of how application secrets like callback URLs are required.
- Good security is expensive. In the last year we have seen API vulnerability on behalf of the biggest companies with the biggest budgets affecting the very best people. These are companies like Google, Facebook, Instagram, Pinterest, and of course, Apple. These companies should spend whatever money they need to secure their infrastructures, yet they still have had problems. If they can't do it right, what does that mean for the rest of us?
- HSMs should be used more often. You want to use them for their code potential to keep important secrets locked away where hackers can't get their hands on them. The problem is that they are expensive right now and start-ups are always trying to look for ways to minimize expense as they get off the ground. The tools for OAuth management are very limited today. If you take a look at the Buffer attack it's very difficult for OAuth tooling to get in the way of a hack that has taken place in a very timely basis.
- Analytics is a challenge. How do people know that something bad is happening at the time that it's happening and how quickly can they take whatever measures are necessary, such as OAuth token revocation to stop it. I mentioned earlier that

two-factor authentication, otherwise known as 2FA, doesn't necessarily work with APIs. What do we do about that? We were looking at 2FAs a big potential solution for many of these problems, but not all of them. Is there a way to figure out a way to bridge that gap?

- IoT is also going to be a major challenge. If companies like Google and Facebook have had problems locking down their APIs, what about all these companies with connected devices coming online? How are they going to be secured? We really don't know the answer to that.

- What are we going to do about documenting all of this and disclosing it in a way that the whole industry can move forward? Everyone acknowledges this is an but we still somehow need work together to solve the problems. There are still companies that just get very quiet when they experience a hack, and there's not a lot of sharing. That is just not the way to proceed in the face of these giant problems that everyone has in common.

How to secure your API

You cannot start contemplating an API strategy without bringing in the chief security officer and as many security people you can into the room, because you are essentially creating an interface that people are going to try to attack. Security has to be a forethought, not an afterthought. A lot of companies start creating an API and by the time they are done, they realize they have to secure this. That's the wrong way to go about it. There a a number of tactical things you can do to make sure at the outset that your APIs are secure.

Don't store OAuth tokens in your REST APIs in clear text, and don't store your API secrets in your source code in clear text. It's not the API itself that's the problem, but rather it's the adjacency—it's the things that are around the API that are the most vulnerable. Just because you think you've secured the API itself doesn't mean it's secure. These hackers are very tenacious. They know how to get around that security. You have to think much more holistically about all of the things that are adjacent to your API and how you secure those as well. The cases that I provided demonstrate that in spades because these were not just attacks on the API. They were attacks on things that were adjacent to the API that eventually led them to successful penetration.

One of the questions that arises when dealing with security is how much human error creates security vulnerabilities? It depends on what your definition of human error is. Is it a human error when a developer who's sitting at the computer gets successfully phished and suddenly their credentials are discovered? One human error, certainly with Apple, was the fact that they did not have rate limiting on that one API. That was an oversight. The question is, for each of these types of human errors, if you want to call them that, what do you do to keep those from happening?

Certain organizations will continually test their employees with fake phishing attempts to get them more sensitized to the fact that people will always be trying to obtain their credentials. When you train people to know what to look for when phishing attempts come into their inbox, the likelihood that they'll fall prey to those phishing attempts is greatly reduced.

You look at the rate limits—how was it that Apple has to ask themselves why rate limiting was not put on that particular entry point into their kingdom? What design tool didn't enforce the best practice of having rate limiting on every API? Where is is that checklist of things that every single API has to perform? In any organization, for the people who are building, developing and designing our APIs how do we make sure that everybody has to check off that list and make sure that all of the best practices, including applying rate limiting, are applied to those APIs.

Some of this can be enforced through design tools. For example, there are design tools out there in the marketplace where if you come up with a pattern for the security you want to enforce, you can at least repeat that pattern across all of your APIs. There are lots of examples of these best practices that need to be put in place and then from enforced from a policy perspective.

API developers need to secure API credentials with the end application developer. Can these be shared in a safe way? If you are an API provider you're developing APIs for your company and providing APIs to developers of mobile applications. When an application developer comes in and says they want to use that API, they are issued a series of secrets that only you know as the API provider and that developer knows. Those secrets are specific to that developer. This is a good way to know whether these applications have a secure infrastructure. You look at these secrets and say, I know that's Joe's application calling us right now.

One of the problems happens when those secrets are blown open. In fact, when those secrets get discovered, hackers will publish the secrets and put them up for sale on the internet. The question is not only how the API provider keeps those secrets confidential, right, how to make sure the API developer doesn't compromise those secrets. This is weak point in the API's autonomy because often the developers of applications will just very haphazardly expose secrets.

For example, they might have a public code repository on GitHub, and often if you go into it the secrets are exposed in clear text right inside of the service code. Every API provider, going back to this API security first approach, should publish a document that very clearly provides security recommendations to developers to make sure that they are following best practices to keep those API secrets secret. They should not trust developers to just do what's right because many developers don't actually know what to do with those secrets and leave them exposed.

Those security policies should recommend that they use HSMs to encrypt those secrets on a separate infrastructure. Make sure on the API provider side that any of the data is encrypted.

What's next in API security

The Internet Engineering Trade Federation (IETF) has what's called a OAuth working group. That group has produced some API security standards If you look across many solutions that are in the API stage, all sorts of solutions, they support those standards. They are going to improve as a result of the work that's being done and will eventually find their way into many of these solutions. It will be much easier to fix these problems. You won't have to get out your alligator clips and your wires and figure out how to solve it on your own. Some of these problems will be much easier to address because they'll be built into the solutions that people are using.

Another prediction is that we obviously haven't seen the last of these transgressions. Not a month is going to go by where we don't see some other highly scaled hack against an API that will result in some very embarrassing news to the API providers and obviously for customers. I think that this will be very damaging to some companies, and that could potentially be very deleterious to the business. We'll see stories of that nature in the future. As the IoT economy becomes very real we are going to hear a lot more about things that were compromised. The more things that come online the more damage can be done because the companies that make those connected devices have no idea what they are doing in the area of security. This is an area of huge risk and as APIs become the fundamental building block of the entire internet this becomes an issue of internet security. In many of the stories that you see in the mainstream media about these attacks it will not be apparently obvious that an API was involved. It will just be a such and such company hacked, millions of dollars lost, or customers affected in this way, that creates major headaches for everybody.

We just have to hope that everybody who is in a position to help fix the problem works together on it as opposed to working independently of each other. We need to collaborate because it's in all of our best interests for all of us to come up with a solution together. APIs provide a great deal of power to businesses and better experiences for employees, customers and partners. We don't want the advantages that API-led connectivity brings to be derailed by security issues.

CHAPTER 13

Protect your APIs: Best practices for API security

By Nial Darbey, MuleSoft

Executive Summary

APIs have become a strategic necessity for your business. They facilitate agility and innovation. However, the financial incentive associated with this agility is often tempered with the fear of undue exposure of the valuable information that these APIs expose. With data breaches now costing $400M or more[11], senior IT decision makers are right to be concerned about API security.

In this paper we show how MuleSoft's Anypoint Platform can ensure that your API is highly available to respond to clients and can guarantee the integrity and confidentiality of the information it processes. We explore in depth the main security concerns and look at how the IT industry has responded to those concerns. We present Anypoint Platform as fully capable of managing and hosting APIs that are secure according to the highest industry standards.

Introduction

A secure API is one that can guarantee the confidentiality of the information it processes by making it visible only to the users, apps and servers that are authorized to consume it. Likewise it must be able to guarantee the integrity of the information it receives from the clients and servers with which it collaborates, so that it will only process such

---

[11] Verizon Enterprise Solutions. 2015 Data Breach Investigations Report.
http://www.verizonenterprise.com/DBIR/2015/

information if it knows that it has not been modified by a third party. In order to guarantee these two security qualities, the ability to identify the calling systems and their end-users is a prerequisite. What we have stated also applies to those calls that the API makes to third party server. An API must never lose information so it must be available to handle requests and process them in a reliable fashion.

In this chapter we use the term API in a broad sense to include both the interface definition and the service or microservice which implements it. We recognize that many of the standards and examples we present are oriented towards HTTP, but with our broad definition of the term API, we also envision the use of event-driven approaches with message brokers.

We also utilize the terms: users, apps, clients and servers. users interact with apps (application software) which are clients to your API. In contrast, your API acts as a server to the app. APIs can also act as clients to other APIs, Web services, databases etc., all of which we refer to as servers. It is common practice to use the term messaging to describe API calls. We utilize both expressions interchangeably here.

Section 1: Identity

Identity is core to the world of security. You must be able to recognize the apps that consume your API, the users of the same and the servers that your API calls out to. Likewise, your API should be able to identify itself to both apps and servers.

You need an identity store to which you can refer to verify your recognition of apps and users. The identity store could be a database, but an LDAP server is the most popular solution. Active Directory is a popular LDAP implementation. In an LDAP server you typically store usernames, passwords, digital certificates, some personal details and the organization groups to which users belong. App IDs can also be stored here.

An identity provider is software which is dedicated to managing the interaction with the identity store(s) for authentication and authorization purposes. Your API can function in this role though it is much more preferable to delegate this responsibility to the identity provider.

1.1 User and app authentication

When you are presented with an app ID or a user's username (claim) in a call to your API, you must be able to verify the authenticity of the claim. This is done with some form of a shared secret. When your API acts as Identity Provider, it typically authenticates the claim by passing the same credentials to the LDAP server.

1.1.1 Username and password credentials

This is the simplest form of authentication. When it is exposed to users, it places the burden of memorizing the password on them. When it is realized with system-to-system authentication, then a password to a server may end up being shared by multiple other APIs.
The use of username / password pairs as credentials is a very common practice but it is not recommended from two perspectives:

Passwords have a level of predictability whereas the ideal is to maximize on randomness or entropy. Username / password pairs are a low entropy form of authentication.

The maintenance of passwords is difficult. If you need to change a password then you immediately affect all apps that make use of that password. Until each of these has been reconfigured you have broken communication with them. As a consequence there is no way you can block access to one app in particular without blocking all the apps that use the same username and password.

1.1.2 Multi-factor authentication

Recognizing the weakness of username and password credentials, an app using Multi-factor Authentication (MFA) demands from the user a one-time usage token she receives after authenticating with her credentials. This token may be delivered to her through SMS when the app requests a Multi-factor Authentication (MFA) Provider to do so. The user may also have a digital key which provides her with a token that the app can validate. An RSA SecurID is an example of this. When the app receives the token which it validates with the MFA provider it proceeds to consume your API.

1.1.3 Token based credentials

Token based credentials are a better alternative to username password credentials, which provide higher entropy and a more secure form of authentication and authorization. The idea is for the Identity provider to issue tokens based on an initial authentication request with username / password credentials. From then on the app only has to send the token, so the net result is a great reduction in username / password credentials going to and fro over the network. Also, tokens are usually issued with an expiration period and can be revoked. Furthermore, because they are issued uniquely to each app, when you choose to revoke a particular token or if it expires, all the other apps can continue to use their tokens independently.

In Figure 3, Janet signs into her app. The app authenticates her and requests a token from the Identity provider. This authenticates her with the identity store and then responds to the app with a token. The app proceeds to call the API with the token.

1.2 API and server authentication

Your API must be able to authenticate itself to the apps which consume it. Likewise when your API interacts with servers, they must authenticate themselves to the API. In both cases you strive to avoid man-in-the-middle attacks which sometimes take the form of malicious software pretending to be a server or indeed your API.
We will take a look at the typical form of authentication for these scenarios when we address public key cryptography in section 2.2.1

1.3 User and app authorization

1.3.1 Role-based access control

Typically, every business, enterprise or organization is divided into groups of employees around related business functions. For example, think of the nursing team and the medical doctor team and the catering team in a hospital. The employees working within an organization have a static function defined by these group boundaries.

This group information can be used when software users interact with an app and you need to restrict their access according to the authorization or access control rule in place for that software. You can use the group they belong to in order to identify their role when using the app. Groups are role and app agnostic. They are purely business level divisions. LDAP servers use the concept of groups for this purpose. Identity providers are responsible for retrieving this group information from the Identity store. A role is an app specific definition of access control. A user will typically adopt multiple roles defined by each app she uses.

Role-based access control (RBAC) represents a very simple access control mechanism. An app need not keep a record of each user's level of access to its functions and data. Rather it uses roles to abstract away from those details and assign degrees of access to groups of users that the role represents.

1.3.2 Attribute-based access control

Going beyond the static assignment of roles to users based on the organizational groups to which they belong, Attribute-based access control (ABAC) aims to facilitate the dynamic determination of access control based on some sort of circumstantial information available at the time of the API call.

Things like the time of day, the role, the location of the API, the location of the app and combinations of conditions contribute to the determination of the degree of access. XACML is a standard which defines the rules that must be executed in order to evaluate the level of access at the time of the API call. The understanding is that this may change from call to call. ABAC often dictates the requirement that your API will respond with subsets of data according to the access control decisions related to the user.

1.3.3 Delegated access control with OAuth 2.0

The HTTP based OAuth 2.0 framework allows an app to obtain access to a resource exposed by your API either on its own behalf or on behalf of the user who owns the resource. Thus it allows users to delegate access control to third party apps.

To facilitate this, your API must collaborate with an OAuth 2.0 authorization server, checking each incoming call for an access token which it must validate with the authorization server. The response from the authorization server will indicate whether the access token is valid (it was issued by the OAuth provider and it hasn't expired) as well as the scope of access for which the token was issued.

1.4 Federated identity

The token-based approach to authentication allows for the separation of the issuing of tokens from their validation and thus facilitates the centralization of identity management. The developer of each API need only concern herself with incorporating validation logic within the API so that upon invocation, it looks for the token in the request and then validates it with the centralized identity provider. If the token is deemed to be valid (the user or app to whom the token was issued have sufficient authorization for this call), then the API should proceed to process the call. The parties involved in this call form part of a single security context. In other words, the Identity Provider is able to recognize and authenticate the app, the user and the API because their identity and shared secret password are in its identity store.

However, your API will not always be exposed to an app and user which the Identity provider can possibly recognize. What if you wish to expose your API to an app in the hands of a user from another company or even from another Business Unit within your own company? Large companies can have many security contexts with separate Identity stores and providers. Federated Identity solves this problem and federated Identity providers collaborate to facilitate the authentication and authorization of users who belong to different security contexts.

In Figure 4, Janet signs into her App. The App authenticates her and requests a token from the orders identity provider. This authenticates her with the orders Identity Store and then responds to the App with a token. The App proceeds to call either the orders API or the shipping API. In both cases the orders identity provider validates the token. The orders identity provider and the shipping identity provider are federated. The shipping identity provider knows that the token has been signed by the orders identity provider to which it delegates the token validation.

1.4.1 Single sign-on multi-experience
The Security Assertion Markup Language (SAML) is an industry standard which has become a defacto standard for Enterprise level Identity Federation. It allows identity providers to communicate authentication and authorization information about users to service Providers in a standard way.

A SAML assertion can be issued by an identity provider in one security context and be inherently understandable by an identity provider in another context. SAML assertions typically convey information about the user including the organizational groups to which the user belongs, together with the expiry period of the assertion. No password information is provided. The identity provider which issues the assertion signs it. The identity provider which has to validate the assertion must have a trust relationship with the issuing identity provider (see digital signatures 2.1.1).

The primary driver for the use of SAML for use within the enterprise is Single sign-on (SSO). Users don't have to keep separate identities for every application software that they use. Rather they sign on once with an identity provider and from then on any links to applications allow them to bypass the login page of each of these. Such a setup ultimately delivers the desired user experience of not having to maintain multiple sets of username and password credentials and of signing in once and subsequently bypassing login pages to all of the applications within the enterprise. This SSO experience is usually delivered with a UI portal which has links to all relevant applications that eliminate the need for any further authentication by the user.

SAML can also be used within the context of APIs. We explore this next.

1.4.2 Single sign-on single experience
The expectation that an identity provider in one security context will understand a token issued by an identity provider in another security context may very well be reasonable within the boundaries of a particular Enter-

prise. However, this may not be so for external company acquisition scenarios and for dealings with partners and SaaS. The industrial standards which we will explore next facilitate the necessary interoperability which allows for any API to deliver its service to apps while relying upon a federated identity provider for the authentication and authorization of the users which are outside of its security context.

1.4.2.2 WS-Security with SAML assertions

WS-Security (in particular WS-Trust) allows identity providers to expose SOAP Web Services that will issue identity tokens to requesting Apps. A SAML Assertion is one such possible token. The same app can then invoke a SOAP Web Service with the SAML assertion in the header.

WS-Security also caters for the general needs of integrity and confidentiality through XML Signature and XML Encryption (see message integrity and message confidentiality below).

1.4.2.2 OpenID Connect with JWT ID tokens

OpenID Connect is built on top of OAuth 2.0 to provide a federated identity mechanism that allows you to secure your API in a way similar to what you would get were you to exploit WS-Security with SAML. It was designed to support native and mobile apps while also catering for the enterprise federation cases. It is an attractive and much more lightweight approach to achieving SSO within the Enterprise than the corresponding WS-Security with SAML. Its simple JSON / REST based protocol has resulted in its accelerating adoption.

Apart from OAuth 2.0 access tokens, OpenID Connect uses JWT (jot) ID tokens, which contain information about the authenticated User in a standardized format. Your API can make an access control decision by calling out to a userInfo endpoint on the identity provider to verify if the user pertains to a certain role. Just like SAML assertions, JWT ID tokens are digitally signed (see Digital Signatures 2.1.1) so a federated identity provider can decide to accept them based on its trust relationship with the identity provider that issued them.

Section 2: Confidentiality, integrity, and availability

2.1 Message integrity

Message integrity goes beyond the authentication of the app and the user and includes the verification that the message was not compromised mid-flight by a malicious third party. In other words, the message received by your API is verified as being exactly the one sent by the app. The same goes for when your API acts as client to a server.

2.1.1 Digital signatures

We humans sign all kinds of documents when it matters in the civil, legal and even personal transactions in which we partake. It is a mechanism we use to record the authenticity of the transaction. The digital world mimics this with its use of digital signatures. The idea is for the app to produce a signature by using some algorithm and a secret code. Your API should apply the same algorithm with a secret code to produce its own signature and compare the incoming signature against this. If the two match, the API has effectively completed authentication by guaranteeing not only that this message was sent by a known app (only a known app could have produced a recognizable signature), but that it has maintained its integrity because it was not modified by a third party while in transit. As an added benefit for when it matters with third party apps, the mechanism also brings non-repudiation into the equation because neither the app, nor the User can claim not to have sent the signed message.

2.1.2 Message safety

Even when you know that your API has been invoked by an authenticated app and user and the message has arrived with its integrity guaranteed, you still need to protect against any potentially harmful data in the request. These attacks often come in the form of huge XML documents with multiple levels of nested elements. JSON documents may also contain huge nested objects and arrays.

2.2 Message confidentiality

It is all very well to rest assured with the integrity of a message sent by a known app, but the journey from app to API may have been witnessed by some unwelcome spies who got to see all of those potentially very private details inside the message! Thus, it is necessary to hide those details from the point of delivery by the app to the reception by the Server. An agreement is needed between the app and API in order to be able to hide the details of the message in a way that allows only the API to uncover them and vice versa.

2.2.1 Public key cryptography
The age old practice of cryptography has made a science of the art of hiding things! IT has adopted this science and can produce an encryption of the message which is practically impossible to decrypt without a corresponding key to do so. It is as if the client had the ability to lock a message inside some imaginary box with a special key, hiding it from prying eyes until the server unlocks the box with its own special key. Digital signing discussed above produces signatures in this very way. Cryptography comes in two forms: symmetric, when both client and Server share the same key to encrypt and decrypt the message; and asymmetric, when the Server issues a public key to the client allowing the client to encrypt the message, but keeps a private key which is the only one that can decrypt the message: one key to lock the message and another key to unlock it!

2.2.2 Digital certificates
Digital certificates are a means to facilitate the secure transport-level communication (TLS) between a client and a server over a network in such a way that the server can authenticate itself to the client. This is made possible because the certificate binds information about the server with information about the business which owns the server and the certificate is digitally signed by a certificate authority which the client trusts.

2.2.3 Mutual authentication with digital certificates
In most cases it is the server which authenticates itself with the client. However, there are also scenarios in which the server demands the authentication of the client. The server requests the client certificate during the TLS handshake over the network. One thing to keep in mind is that the server controls whether client authentication occurs; a client cannot ask to be authenticated.

Mutual authentication with TLS certificates is ideally suited to the type of system to system communication that you see when your API acts as a client to a server, whether the server be another API or a database or any other system entity. Missing from this sort of communication is the human User. Hence, the security credentials exchanged between the two parties are far easier to manage.

2.2.4 HTTPS

By utilizing TLS, your API can expose itself over HTTPS and guarantee both Message integrity and confidentiality at the same time. Public Keys are emitted on certificates which have been digitally signed by independent and trusted certificate authorities, thus guaranteeing that the public key was issued by your API. Once the initial handshake has been completed with the app by the exchange of messages using public and private keys, the communication switches to the more efficient symmetric form using a shared key generated just for the duration of the communication, all of which occurs transparently.

2.4 API availability

Your API must guarantee that it is always available to respond to calls and that once it begins execution on the call, that it can finish handling the received message right the way through to completion without losing data. This can be achieved by horizontally scaling the API across multiple servers and by handing off the processing of the message to a message broker which will hold the message till the API has completed its processing. The understanding in this latter scenario is that another process is subscribed to this message publication and thus continues the processing asynchronously.

Thus it is clear that reliability is a step beyond mere availability. While an API may (through horizontal scale-out) be available to respond to all calls because a load-balancer in front of the API can propagate calls to the correct hosting server when any of the other servers are down, this still may not be enough because the API may fail mid-way through processing. In a reliable architecture, the API would receive the call and then leave a message on a message-broker queue (JMS or AMQP for example). Even when the service which has subscribed to the queue is down, the broker can hold onto the message for later consumption when the service comes back online again.

Section 3: Mule runtime security capabilities

The Mule runtime addresses a broad set of security concerns with best practice solutions for transport and message level security. Your API can be hosted by Mule, exposed over HTTPS and can facilitate TLS client authentication. Likewise, your API can make calls out to servers over HTTPS and issue client certificates as needed. Keystores and truststores are used to store TLS certificates for these scenarios. Messages that are sent to exchanges and queues on Anypoint MQ, MuleSoft's cloud messaging solution, can be encrypted. The Mule runtime can also encrypt and decrypt messages, digitally sign them and verify the validity of incoming digital signatures. IP white and black-listing is also available.

With these capabilities, the Mule runtime addresses the concerns of exposing highly available APIs that authenticate and authorize incoming calls while guaranteeing message integrity and confidentiality.

3.1 Message confidentiality on the Mule runtime

3.1.1 Mule HTTPS Connector
An HTTPS listener can be configured with reference to a keystore so that your API can authenticate itself to the app. When client authentication is demanded from the app, the listener can reference a truststore. Similarly, when your API needs to interact with a server over HTTPS, you can use the HTTPS Request Connector which references a keystore to authenticate itself and a truststore for when digital certificates are not recognizable by the standard Java JDK truststore (cacerts).

3.1.2 Mule Encryption Processor
The Mule Message Encryption Processor can change the content of a message so that it becomes unreadable by unauthorized entities. Mule can encrypt the entire payload of a message or specific parts of the payload, according to security requirements, using different encryption strategies.

3.1.3 Dynamic data filtering with DataWeave
For ABAC scenarios (see Attribute Based Access Control 1.3.2) you will need to filter the payloads that your API sends back to apps based on

the degree of access determined either for the app or for the user. DataWeave is MuleSoft's data transformation engine which transforms between different mime-types using a simple expression language which is common across all data formats. The language can be used to remove and / or mask data fields in the payload, whatever the structure of the payload and the location of the field. The same expression can be stored in a datastore and 'blindly' executed by the engine at runtime. In this way you can cater for dynamic transformation logic that you may not necessarily be able to configure at design time. Rather, at runtime based on criteria decided at runtime, you can choose a transformation and apply it to the payload.

3.2 Message integrity on the Mule runtime

Exposing your API over HTTPS guarantees that it has not been modified in transit. However, authentication and authorization of the request still need to take place.

3.2.1 Mule Security Manager

Central to authentication in Mule is Mule Security Manager. This is the bridge between a standard mule configuration and Spring Security beans.

Figure 9 illustrates how credentials are passed and validated in the solution. The security-manager, as you can see above, delegates to the authentication-manager. The authentication-manager uses the authentication-provider to authenticate the set of credentials. The authentication-provider abstracts away from the details of the system used to do the authentication, whether it be in-memory, LDAP or DB based. The Spring LDAP Authentication Provider uses the BindAuthenticator in order to build a DN based on the credential username with which to bind directly to the LDAP server.

3.2.2 Mule Secure Token Service OAuth 2.0 Provider

Mule can act as an OAuth2 Provider, issuing tokens to registered Apps, applying expiration periods to these tokens and associating them to user roles and fine-grained access control known in the OAuth world as scopes. Refresh tokens can also be issued and tokens can be invalidated. Mule can subsequently validate incoming tokens against expiration periods, roles and scopes and thus grant or deny access to the flows in the Application. Scopes represent broad levels of access to the Mule flows. The provided access token must be sent in with each request and can be validated by Mule to ensure it has not expired or been revoked and that it has the scopes that correspond to a particular flow.

More fine-grained control can also be applied by comparing the role of the user for whom the token was issued with the allowed roles for the flow. The validate filter has a resourceOwnerRoles attribute to specify these. The granularity of access control can be in either the grant or the role.

As we extend APIs outside of our organization we may have to cater for applications belonging to partners. Imagine we were to dynamically expose access to your API to a mobile application. We need only register this new client in your OAuth 2.0 Provider configuration.

3.2.3 Mule Digital Signature Processor
Mule Digital Signature Processor adds a digital signature to a message payload, or part of the payload, to prove the identity of the message's sender. Mule can also verify a signature on a message it receives to confirm the authenticity of the Message's sender.

3.2.4 Mule Credentials Vault
Mule Credentials Vault is for the encryption of properties that are referred to and decrypted by the Mule application at deployment time. These properties are encrypted with a variety of algorithms and are completely hidden from anyone who does not have the key to the Credentials Vault. At deployment time, the key is passed to Mule as a system property. This key should only be in the hands of authorized personnel.

3.3 API availability with Mule HA Clusters

A single Mule server hosting your API is not enough to facilitate high availability. To achieve this, you need to host the same API on more than one Mule runtime. With a load balancer in front of the API, you can guarantee that the API will always handle incoming requests as the load balancer chooses between those instances which are healthy.

Reliability on Mule can be achieved by clustering two or more instances of Mule together which is easy using Anypoint Runtime Manager. In this scenario, we configure the Mule VM endpoint as a reliable handoff mechanism immediately after receiving the message. Another flow processes the message from that same VM endpoint. If the Mule node which receives the message from the VM goes down, then another Mule node on the same cluster will pick up the same Message. For reliable processing of messages between multiple APIs you can use Anypoint MQ (see Anypoint MQ 4.2).

Section 4: Anypoint Platform security capabilities

4.1 Anypoint Platform API solution

Anypoint Platform API solution is a fully multi-tenant application running on top of Amazon Web Services (AWS) and inside a VPC (cloud VPN). Data, metrics, and metadata cannot be accessed across organizations.

Although Anypoint Platform can manage and enforce the runtime security of your API, the API itself remains wherever you have it deployed. Only the configuration of the policies, metadata about your API and analytics about the usage of your API is stored in Anypoint Platform.

4.1.1 API adaptability through policies on API Manager

API management is a discipline which addresses the need to publish your API for consumption by known apps, registering those apps and provisioning them with their own ID and secret identifiers, authorizing the apps to consume your API and adapting the APIs to the potentially different security requirements across the apps.

The adaptability is addressed with what we call policies. These are encapsulations of the types of logic that usually recur across your APIs. Similar to how aspect oriented programming worked, these logical bundles can be applied to or removed from running APIs without affecting their lifecycle. Security is a prime example of such logic. We explore security policies next.

4.1.2 Secure communication between the Mule runtime and API Manager

Using the Mule runtime as an API gateway you can host your API. The Mule runtime communicates constantly with Anypoint API Manager to retrieve policies and report back analytical information about the usage of your API. This communication is initiated by the Mule runtime which authenticates itself with OAuth 2.0 client credentials. You configure the Mule runtime with a client ID and secret which is configured for your particular organization (or business group) in Anypoint Platform. The ID and secret are used by the Mule runtime to get an OAuth 2.0 token to be used in subsequent calls. All calls are to a RESTful service which is accessible over HTTPS. The Mule runtime is insulated from external network outages since it stores a local cache and can continue to operate even if the Anypoint Platform were to become unavailable. Regardless, MuleSoft maintains an SLA of 99.99% for Anypoint Platform API solution. Anypoint Platform is certified via WhiteHat Sentinel.

4.1.3 Security policies

We must return to our discussion about identity in the light of what Anypoint Platform has to offer in its suite of policies. Some of the following policies are inherently dependent on a mechanism to verify incoming identity tokens. All of them address security concerns:

- **Client ID enforcement**: locks down your API for consumption only by a set of known clients.
- **SLA-based rate limiting**: provides different quality of service contracts to your known clients, 10 calls a minute for some, 100 calls a second for others, etc.
- **SLA-based throttling**: same as rate limiting only exceeded calls are queued for next time window.
- **Mule OAuth 2.0 access token enforcement**: validates incoming tokens previously issued by Anypoint OAuth provider upon receipt of client ID and secret.
- **External access token enforcement**: validates incoming tokens previously issued by PingFederate or Open AM OAuth provider upon receipt of client ID and secret.
- **Cross-origin resource sharing (CORS)**: permits your API to be invoked by a JavaScript client (an Angular app, for example) which is hosted on a domain different to your API.

- **HTTP Basic authentication:** authenticates using credentials which are configured in a security manager underlying this policy.
- **IP blacklist and whitelist:** denies or permits calls from only from IP addresses present in this list.
- **Json and Xml threat Protection:** guarantee the safety of the messages passed to your API.

The Mule runtime stores the client IDs and secrets of consuming apps in an Identity Store. When you register a new consuming app in the API Portal, Anypoint Platform generates a new app ID and secret and persists it. Later, when any identity related policy, like SLA based throttling, is applied to your API, the Mule runtime downloads the policy and also downloads the ID and secret for every consuming app registered to consume your API. Thus, when the Mule runtime injects the policy configuration into your API, it also provides access to a local embedded database of IDs and secrets which the policy consults to verify the identity of the calling app. When you choose to integrate your Anypoint Organization with an external identity management technology like PingFederate, this assumes the role of administering and persisting app ID and secrets.

4.1.4 Custom security policies

Policies on the Anypoint Platform are snippets of Mule configuration. As such, custom policies are very easily configurable and can be surfaced on the API Manager portal as siblings to our out-of-the-box policies.

We have a number of custom policies published to Anypoint Exchange. These cover SAML based security use-cases such as the ability to validate incoming assertions, or username tokens.

You can write your own policy to cover any area of logic that is pertinent to your API. For the ABAC scenarios that we described above, you might consider configuring a custom policy which accepts either a full DataWeave expression in the API Manager portal UI at policy application time. This can be executed at runtime on the response payload (See Dynamic Data Filtering with DataWeave 3.1.4).

If you wish to protect your API with OpenID Connect, you should consider writing a custom policy to validate incoming tokens against the authorization server.

4.2 Anypoint MQ

This is a multi-tenant cloud messaging service offering persistent data storage across multiple data centers, ensuring that it can handle data center outages and have full disaster recovery. For compliance with your data at rest policies, Anypoint MQ allows you to encrypt all messages that arrive in either an exchange or a queue.

4.3 Anypoint Platform Virtual Private Cloud

Mule applications can be deployed either to your on-premises Mule runtime or to our fully hosted and fully managed iPaaS. In most scenarios, Mule applications deployed to the iPaaS will need to integrate with systems in your datacenter. In some cases a hybrid architecture is adopted where Mule applications deployed to the iPaaS must integrate with Mule applications deployed to Mule runtime on-premises. Either way there is a need to establish a secure network between the cloud and your datacenter.

Virtual Private Cloud (VPC) enables you to connect your organization in Anypoint Platform to your corporate data centers – whether on-premises or in other clouds – as if they were all part of a single, private secured network. You can configure these networks at hardware or software levels.

VPC can be configured to use IPSec, TLS (over OpenVPN) or Amazon VPC peering to connect to your on-premises data centers. IPsec connections can be configured at the hardware level in addition to a software client. If you already use Amazon, you should use VPC peering. Otherwise, IPsec is in general the recommended solution for VPC to on-premises connectivity. It provides a standardized, secure way to make connections and integrates well with existing IT infrastructure such as routers and appliances.

4.4 Anypoint Platform user roles and permissions

In Anypoint Platform, users belong to an organization and have a set of roles and permissions. API versions and deployment environments are grouped under organizations (and optionally under business groups too), to access them you need to have an account that that owns the necessary permissions and that belongs to its corresponding organization – and to the business group if the resource exists in one.

Roles and permissions can be granted for accessing resources that exist in the master organization, or for resources that exist within a business group. A user that owns any role of a business group is implicitly granted membership in the business group.

Each role contains a list of permissions that define what a user that holds that role can do with the specific resources it scopes. Permissions can also be added at an individual user level, without the need for roles. There are two different types of permissions: those that are for API versions and those that are for iPaaS environments.

Keep in mind that API permissions are API version specific and iPaaS permissions are environment specific – they grant you the ability to do something within a particular API version / environment, not the entire organization. The only exceptions to that rule are the roles API versions owner – which grants ownership of all APIs and all of their versions within the corresponding business group – and portals viewer – which grants viewing access to all portals in the corresponding business group.

4.4.1 Federated user access to Anypoint Platform

Anypoint Platform can be integrated with your organization's external federated identity provider. Opting to use federated identity management for Anypoint Platform gives your users single sign-on access and facilitates OAuth security for APIs using the same identity management system.

Anypoint Platform supports SAML 2.0 identity providers for user management, the following ones were successfully tested working with the platform:

- PingFederate
- OpenAM
- Shibboleth
- Okta
- ADFS

You can set up your Anypoint Platform organization so that when a SAML user belongs to certain groups, it will automatically grant certain equivalent roles in the your Anypoint Platform organization.

Section 5: Anypoint Platform security compliance

When Anypoint API Manager manages APIs from the cloud, it stores only metadata about the APIs and the apps which consume them. The APIs can be deployed on the Mule runtime either on-premises or in our fully hosted, fully managed iPaaS solution. The extent to which Anypoint Platform platform is compliant with external audits is detailed in the whitepaper "Anypoint Platform Cloud Security and Compliance". We summarize them here:

- **FIPS 140-2:** the US government security standard that requires that compliant parties use only cryptographic security modules that have been certified by NIST. Mule can be configured to run in a FIPS 140-2 certified environment. Note that Mule does not run in FIPS security mode by default. There are two requirements:

    Have a certified cryptography module installed in your Java environment

    Adjust Mule settings to run in FIPS security mode
- **SSAE-16 SOC 2:** a regular audit report that states whether or not we are compliant with the policies we give our customers. We get these every January.
- **PCI DSS Level-1:** PCI compliance is a set of requirements designed to ensure that all companies that process, store or transmit credit card information maintain a secure environment. We complete an onsite assessment every September.
- **HiTrust:** a common security framework (CSF) designed to simplify compliance with technical controls derived from HIPAA/HITECH. Put another way, it is a way for a service

provider like MuleSoft to show compliance with the relevant aspects of HIPAA/HITECH to its clients. MuleSoft completes an onsite audit every two years.

Section 6: Summary scenario

Let us consider a hypothetical scenario: Mythical Retail has a chain of stores and delivers an e-commerce solution to their customers. One of their business objectives is to increase their revenue by 20% over the next 18 months. To achieve this goal they aim to improve customer loyalty by providing a compelling omni-channel digital experience. This experience will allow customers to shop with ease wherever and whenever they want, receive appropriate recommendations and offers and see their current loyalty balance in real time.

Mythical Retail wants a 360-degree view of their customers' visits across all touch points. They aim to reward every customer interaction made online and in-store and have a clear view of their customers' spending habits.

Mythical Retail has adopted MuleSoft's API-led connectivity approach to integration and see APIs as strategic assets upon which they can execute digital initiatives. They have invested in clienteling software that can help their sales associates meet and register anonymous customers in-store and identify customers already registered. This software will also provide a mobile point-of-sale (POS) experience. With APIs backboning the clienteling solution as well as the customer's web and mobile interactions, Mythical Retail can guarantee the uniformity of the customer experience across every touch point.

6.1 Securing the customer transactions

Katie is a customer of Mythical Retail and likes to purchase online. She uses the iPhone app to make orders. Katie registered with Mythical Retail through a sales associate in-store. Her details are stored in Active Directory.

This is an overview of the APIs and servers needed to deliver her the capability to place an order on the phone.

1. Katie must sign in to the app.
2. The app must authenticate itself on Katie's behalf and consume the My Shopping API with the relevant degree of access control. Only Katie's data must be accessible for this interaction.
3. All calls between experience, process and system APIs must be protected.
4. The Order Fulfillment API orchestrates the Orders API and the Tokenization API. The latter is exposed by a third-party credit card processing company and delivers obfuscation functionality to Mythical Retail so that 5. Katie's credit card details are never stored in their original form in the systems of record. The call to the Tokenization API needs to be authenticated, signed and encrypted.
5. The Order Fulfillment API creates a business event which it publishes to an encrypted exchange on Anypoint MQ. The published data must be encrypted and the publication call must be signed, encrypted and authorized.
6. The Customer API subscribes to the event on Anypoint MQ and is responsible for increasing Katie's loyalty points. The Recommendations API is also registered to consume this event and gathers the details of the order to feed into future recommendations accordingly. Likewise the Payment API subscribes to the event and finalizes the financial transaction. The shipping of orders is the responsibility of Mythical Retail's partner. Their Shipping API also subscribes to the same event. All of these subscriptions to Anypoint MQ must be signed, encrypted and authorized.
7. The iphone app utilizes the Order Tracking API, which is exposed by Mythical Retail's partner. The partner forms a separate security context and Katie is not in their Identity Store. Her claim must be recognizable in the shipping context.
8. Interaction with the systems of record must be secured according to the requirements of each server.

6.2 Anypoint Platform as part of the security fabric for Mythical Retail

This is an overview of the APIs used to deliver the loyalty experience to Mythical Retail's customers. All the APIs are deployed to **MuleSoft's iPaaS** and managed by **Anypoint API Manager**. **Anypoint MQ** is used for messaging between the APIs and PingFederate is used as Identity provider, MFA provider and OpenID Connect / OAuth Provider. All HTTP APIs in this context are protected with **HTTPS** and **Anypoint security policies** are applied to each of the APIs using **Anypoint API Manager**.

1. Katie signs in to her iPhone app.
2. Part of the digital solution catering to Katie's experience is the My Shopping API which delivers all the relevant capabilities to her iPhone App. This API is protected with the **PingFederate access token enforcement policy**. In order to consume the My Shopping API, the iPhone app must interact with the PingFederate Authorization Server to request an OpenID Connect token on her behalf. PingFederate authenticates her credentials against Active Directory. Upon successful authentication, it generates a token and signs it (see digital signatures 2.1.1) before responding to the iPhone app with the token. The app presents this token in a custom HTTP header on every subsequent call to My Shopping API.
3. All calls between experience, process and system APIs are protected with the **client ID enforcement policy**. Each API has an ID and secret stored in **Mule Credentials Vault**.
4. The calls to the Tokenization API are protected with **client cert authentication** and the payload passed to the API is encrypted and signed with **Mule Encryption Processor and Mule Digital Signature Processor** respectively before sending.
5. **Publications to Anypoint MQ** are protected with OAuth 2.0 and HTTPS.
6. **Subscriptions to Anypoint MQ** are protected with OAuth 2.0 and HTTPS.
7. There is a federated trust between the Identity Providers in both security contexts for Mythical Retail and its shipping partner. Both the sales associate's App and Katie's app can call the Order Tracking API with the access token that they

received from PingFederate. The Order Tracking API validates the token with OpenAM, the identity provider of the shipping company. This is able to verify that the token was signed by the identity provider of Mythical Retail, which it trusts. The Order Tracking API accepts the invocation by the iPhone app and responds accordingly.

8. Sales associates use multi-factor authentication to sign in to their app. The app calls PingFederate OAuth 2.0 authorization server to get a token which it passes to the Shop Assistant API.

9. Interaction with the systems of record is secured in various forms according to the requirements of each server. Tokens and username and password credentials are stored in **Mule Credentials Vault.**

In conclusion

APIs are a strategic necessity to give your business the agility and speed needed to succeed in today's business environment. But with the increasing cost of security breaches, senior IT decision makers quite rightly want assurances that exposing their data via APIs will not create undue risk. Anypoint Platform's features provide the necessary assurances for the confidentiality, reliability and availability of APIs designed and managed on the platform.

CHAPTER 14

Microservices: Self-serving new experiences

by Uri Sarid, MuleSoft

One of the very hot digital trends right now is the notion of microservices. The concept has been around for a little while now, and it's getting a lot of excitement. Microservices is not about one specific technology or another, and it is not really about turning everybody in your company into a 24/7 DevOps team. The idea behind microservices goes hand in hand with both API-led connectivity and the notion that technology development should be decentralized throughout the organization, not centralized in IT. Microservices allow the business to self-serve new technology projects and initiatives based on their own areas of expertise. This is accomplished by picking small teams and giving them a well defined mission, saying, "You own this. Run as fast and as hard and as smart with this particular thing, and then make it work together with all of the other microservices that are happening."

The question that any business that wants to use a microservices approach has to ask themselves is: How can these hundreds and thousands of microservices and these teams actually align, and not lead into absolute anarchy? How do you actually make sure that you have some governance in place? The answer is to look to the the interfaces between them. It's a simple matter of geometry.

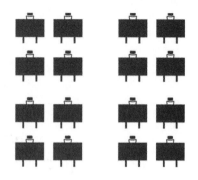

If you've got some services, and you decompose them into smaller services - microservices - they have the same amount of code as the services that they came from, but there are a lot more APIs once you break them up. Therefore, APIs become even more important. A key consideration for business when thinking about microservices is how these teams and how these services talk to each other.

Another important consideration is what you actually want to achieve with this architectural approach. Microservices not only validate a service-oriented approach but are also a way for that approach to be implemented, by taking the need for well-defined services and reusability to an extreme. This approach highlights the need for governance and management; successful implementation must also consider non-technology factors such as development processes and methodologies.

As you read this, you might be thinking, "I know where I want to get to. I want to become more agile. I want my business to be more agile. I want to be more composable. I want to figure out how to turn my business into something that works faster, and I need to keep robustness. I need to keep the tires from falling off. I need to make sure that certain regulations are enforced." It needs to be robust as well as composable and agile and fast moving. All that is great, but then the hard questions must be asked: "How do I actually make this happen? What are the steps? How do I get there?"

Composability requires you to connect systems, and reconnect systems, and then reconnecting them again and changing them around over and over again. Because those connections and reconnections set the clock speed for your business, you also want to make sure that you're good at this connectivity. The secret to being good at this connectivity is to concentrate on the API.

I like to think about APIs in two ways - on the outside, and on the inside. APIs on the outside are easier to imagine. Google has an API and Uber has one, and Salesforce drives most of their business through their APIs. The API economy consists of a number of services that you can pick from and connect for best-of-breed functionality. For billing, there's Stripe; for transportation, there's Uber, and for geolocation there's Google. These external services can be put together and used in various places in the business at various times; they can be recomposed and mixed and matched. But the real power of the composable enterprise, that a connected approach with microservices architecture can bring, is that a company can be that composable on the inside as well. Inside your company there are small teams, and each one can offer one capability as an API, and those capabilities can be composed on top of other teams, and they could choose whether to compose on top of their own teams or on top of external APIs. Everything can be that dynamic, that composable, and that well defined.

To illustrate this, here are some simple scenarios that are very familiar, so we will reimagine them from a microservices perspective.

Scenario 1: Sales and Licensing

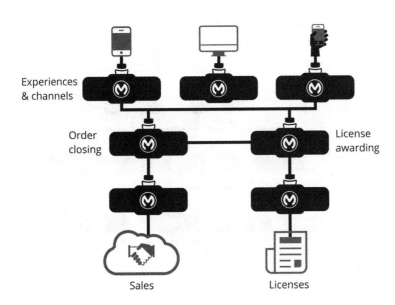

Let's look at a very simple one where there is a sales system where orders are taken, and then there's a licensing system, so that every time an order comes in a license needs to be awarded to that person. An integration would usually be built between them to make one trigger another. Perhaps you could build an integration to connect all sorts of legacy systems or connect an older sales system to a new one in the Cloud? What if you thought about things a little bit differently? How would you like your sales system to actually look from a system perspective?

The licensing system is worth considering too. The code in there may be some ugly thing that somebody wrote a long time ago, so could it be rebuilt as an API that you have some control over? Once you're in control of these APIs, then you should be able to create processes on top of them. For example, an order closing process takes these better APIs for sales and licensing and puts them together in a process that you control and exposes that itself as an API. Similarly, your business processes themselves become exposed as APIs. You are in control of the process, and it's now open to who calls that process, who actually invokes it, which means that on top of that you can start building all sorts of interesting experiences and channels. For example, other people can partner with your business to help you close sales.

Here's another example. Very often through mergers and acquisitions, companies end up with multiple HR systems.

Scenario 2: Multiple HR Systems

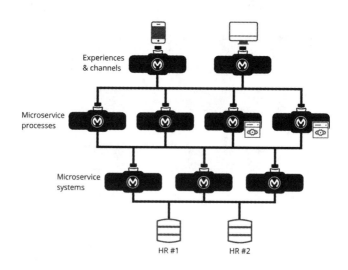

The classic approach to this is to put a façade in front of this that makes the two systems look like there is a single system that takes care of all the nastiness underneath it. If it's rethought with a microservices approach, those systems can not only be cleaner, but something you can leverage more in the future. To rethink those systems, step back and say, "What are the fundamental things they are dealing with?" they are addressing employees, payroll, and benefits, all of which are separate concerns. There could be APIs for each one of them, and those APIs could be mapped to the HR systems, and all the legacy code could go on underneath that abstraction. It's a very clean model on top. There's a team that owns employees and a team that owns payroll and so on. Once the substrate is laid out that way, without reimplementing the systems, but just by putting APIs on top of them, some interesting processes can be built on top of that. You could change and customize your employee onboarding and offboarding process. What training would you like employees to have? What's the best way offboard an employee to make sure they don't have access to systems they should no longer have access to? When there's a life change event or when you want to enroll in some particular thing, how should that actually work?

With a microservices approach, the business can start thinking its processes without having to deal with all of the systems that live underneath that. That means that the HR team, who handles governance, is going to be pretty happy that it is in control of the process, but the employees that are on the receiving end of this are hopefully thrilled. You've put the human back in human resources. It's actually enjoyable and rewarding and something that you can actually be proud of. Again, this is not a rip-and-replace scenario - you've made use of your own legacy investment. There may be some custom logic you have to write, but the main connectivity remains very agile, very composable.

Let's look at a third example, and then we'll draw some conclusions. Imagine a company where there is a need to onboard a lot of content from a variety of content suppliers, some of which were very old, some of which were very modern. They all dropped their stuff in an FTP site, and integration was there, an ingestion pipeline to take that content, sort it through the catalog, convert it into a variety of new formats, and eventually put it in some Cloud repository where mobile applications could access it.

Scenario 3: Content Delivery

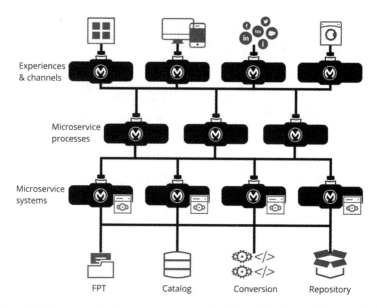

Experiences & channels

Microservice processes

Microservice systems

| FPT | Catalog | Conversion | Repository |

There's a better way to leverage systems like this that allows the business to go a lot faster. Look at these systems and ask the question: "What are the fundamental objects they are trying to deal with?" There are files. These files shouldn't shouldn't be tied down to an FTP system. You're going to have to deal with content metadata, but you may have multiple catalogs. Certain things have to happen on schedule, and certain things have to have entitlements or licenses awarded to them.

Think about all of those objects and functions as individual APIs. The systems underneath don't have to be rewritten. You may have to put some logic in order to make them work the way you want them to, but the end result of that is that you can start building out processes that are very clean and very flexible. This is how content is ingested, this is how it is distributed, this is how we syndicate the fact that there is content, and now other people can help sell it because now you can start rolling out all sorts of interesting experiences. Other people now can supply content to you through a variety of mechanisms that can be adapted on the fly. New digital channels can be leveraged to take that content, and find new revenue streams for it, and do more social media integration, or even integrate with the in-store experience without having to touch all those older systems underneath. The microservices approach allows you to tap into all those capabilities you had before, but not burden everybody who is trying to compose rapidly on top of them.

This leads to the most important component of the microservices approach, which is a validation of our notion of API-led connectivity. It is a self-service machine. It's not about one team learning how to do this. It's about enabling every team in the company to be able to do this kind of thing. For example, some of MuleSoft's customers have teams that are looking at code that literally runs on mainframe; they are currently exposing them as microservices so the rest of the company can benefit from them. There are companies who have ancient applications that are entirely written in database stored procedures—reams of PLSQL going everywhere. That doesn't have to be locked down there, because a team that understands those applications can expose them all as APIs, and then the rest of the company can leverage them. Logistics vendors, who have all sorts of old and new partners, might be wondering, "Are APIs for me?" The answer is absolutely yes because APIs are what allows them to reach out to their old vendors and their new partners and make it all work together. More than 50 percent of interactions between systems inside of an enterprise are asynchronous, but APIs belong there, too. You need a contract in order to be able to tell who is producing information and who is consuming information so you can recouple and recompose all of those things.

In order to make this self-service model as efficient as possible, people need the tools to operate it very efficiently. One of the most important tools to make this approach work is to start from the desired goal and work backwards. An API-designed life cycle is warranted and must start with designing that API. The questions that you must ask are: "What do I want my API to be like? What would I like it to look like?" and not "What systems do I have?".

The API is the fundamental building block to a microservices approach to architecture. It has three pieces. It has managed API capabilities in front, so you know that whenever an API is deployed, it will be governed and will be under control. It has all the logic in the middle to connect things together, and it has got the connectivity on the bottom to reach out to other systems and pull them in. If this is given out to every development team in your organization, that's when the magic happens. All of these APIs attached to microservices know how to talk back to a core set of platform services, which means that as you're building out these APIs, the design is automatically controlled, documentation is automatically published for all the developers, and analytics are automatically built in. You don't have to go back and retrofit everything. All sorts of concerns that developers don't want to have to worry about are built into this platform automatically on top of a bunch of shared services.

With this building block, a pattern architecture is laid out. It's not a reference architecture in the sense that all the pieces have to be built. It's a pattern for how these systems should be considered.

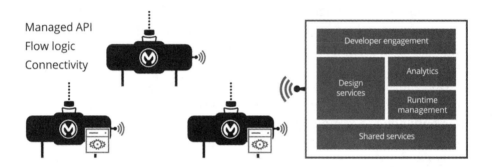

This is what we call carrot-based governance. Instead of being overly prescriptive with teams about how to build, the idea here is make it easier to build things the right way. It's important to ensure that every one of your building blocks actually comes with built-in positive governance.

With that in place, the end result looks something like this.

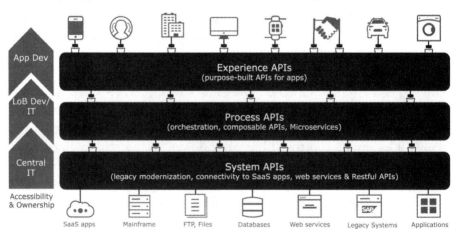

You look at all the systems that you need to connect to and make sure that they actually have APIs on top of them. These are called system APIs. they are your systems of record. they are the things that actually make things happen. Then on top of that, now you can separate out the business processes that you want to do and design them the way that you want to happen and expose those, as well, as process APIs. The third layer on top are the experience APIs, ones that are created to support very specific experiences. There may be lots of these, and they may be very short lived, but since the governance is also happening at lower levels, these APIs won't disrupt your process because they are built on top of reusable APIs.

While this tiered architecture is important, it is the opposite of a big monolith. To get the most out of a microservices architecture approach, a wholesale enterprise transformation is not necessarily the right way to go. It's actually much better to start at the bottom and make this more of a self-serve and distributed model.

Microservices validates our approach to API-led connectivity and extend the notions of reusability and composability in the enterprise. We believe strongly that the way to win in this business environment is to enable the entire business to leverage technology and create great experiences for relevant audiences, which connect back to central data assets through APIs. Microservices is an architectural concept which makes that simple.

CHAPTER 15

The dawn of DevOps

by Shana Pearlman, MuleSoft

It would be very easy to think that the API-led connectivity approach to the modern operating infrastructure is a mere exercise in changing an approach to technology. A little cloud adoption here, a little API development there, and suddenly, every business is ready to compete in the globalized, digital age. However, being able to compete and win in today's hyper-connected, hyper-fast business environment is not just about deploying selected bits of software. It's actually a cultural transformation as well; a shift in how we think about the way people work, rather than just the particular technological tools they are using.

This is why the philosophy of DevOps has become trendy in CIO circles—because it is an approach to changing the business mindset about how technology is created, deployed, and used. CIO.com writer Rich Hein says, "Some refer to [DevOps] as a philosophy. Using both lean and agile methodologies, organizations bring IT operations and development teams as well as quality assurance people together throughout the software life cycle to create a more collaborative process that, in the end, should deliver software and/or services in a faster and more continuous manner. Traditionally, elements of IT have been siloed. DevOps aims to break down those silos to get everyone working towards the same goal."[12]

Like the notion of the composable enterprise, which consists of being available to everyone who wants to use it, and able to be broken down and reassembled to create new experiences, the idea behind the

---

[12] Hein, Rich. How Devops Can Redefine Your IT Strategy. CIO.com, March 17, 2015.

DevOps cultural shift is that same agility and composability applies to the development process itself. Laurent Valadares, the director of cloud operations at MuleSoft says, "We can release new products, and new infrastructure and evolve it, because we take a more modern approach to infrastructure. We can iterate, enhance and build on a program from scratch, without all the traditional steps to go through. We don't need to buy hardware, the QA and steps are all integrated into programmable infrastructure. Typical infrastructures are harder to iterate on. The DevOps process can improve things as they occur, so it's not hard to make radical new changes. We can take the iterative process and move forward faster than the classic three year plan."

Definition of DevOps

Steven Norton, in the Wall Street Journal, notes that "DevOps still defies a single definition and can change depending on the size and type of company where it's used. But it's picking up steam in business circles as a set of practices that can get developers and operations teams collaborating to deliver stable, secure software and respond quickly to user demands. It also can serve to break down siloes and connect developers and operations teams throughout the product life cycle."[13]

Valadares says that this methodology is often adopted by startups and cloud-based software providers, as their infrastructure and business practices aren't as established as other types of companies. But he notes, "some of the more enterprise companies are shifting in this direction, but because they have such investment in this current infrastructure, making the move is harder when you are established. And some companies are conservative. They are going to wait until everyone is 100% sure it's safe."

But the time to wait may be growing short. Gartner predicts that by the end of 2016 DevOps will have moved from a niche approach employed by large cloud providers to a mainstream strategy used by 25% of Global 2000 organizations.[14] And as the impetus to implement DevOps comes from the CIO's office, and can have profound effects on operational efficiency and speed of delivery, a change in strategic approach

---

[13] Norton, Steven. Study: Devops Can Create Competitive Advantage. Wall Street Journal, June 4, 2014
[14] Samuels, Mark. Four CIO Tips for Implementing DevOps. ComputerWeekly.com, November 2015.

might well cement the CIO's role as the Chief Innovation Officer, if the benefits offered by DevOps are realized.

Advantages of DevOps for the business

A recent report[15], conducted by Utah State University and sponsored by Puppet Labs, revealed that DevOps practices can boost IT performance and ultimately improve a business's bottom line. High IT performance correlates with DevOps practices like proactive monitoring and continuous delivery and integration. Interestingly, the report also showed that the longer an organization uses and improves upon DevOps practices, the better it performs, which can lead to improved performance for the entire company.

"This is where DevOps is different. It isn't just IT, it's the practice of IT," says Nicole Forsgren Velasquez, an assistant professor at Utah State University and one of the researchers on the study.

The benefits to DevOps lie in four key areas:

- Reliability
- Security
- Cost-effectiveness
- Agility

One specific benefit that DevOps can provide, says Laurent Valadares, is optimizing processes to improve employee productivity. "DevOps has an enabling function - they have an architectural role with working with the CTO and architects to evolve the platform, which must work alongside corporate IT. One area in which they work together is implementing single sign on processes for internal employee services, which is much easier for employees to handle.

Another example is specific to the finance industry. In financial services, infrastructure costs are a major line item in the budget, so a DevOps approach can allow the IT organization to manage costs on an hourly basis, rather than a classical approach on a yearly basis. You can

---

[15] *Ibid.*

upgrade your capability, retire what you need to, all based on what you need to do when it needs to be done. This makes the business move faster."

Are there any risks to adopting DevOps?

This all sounds like the Holy Grail of business decisions; there isn't an executive in the world who wouldn't sign off to faster delivery of products at a cheaper cost. But, says Valadares, there are a couple of things to consider before making a wholesale adoption to DevOps.

"There are some security risks you need to plan for," he says. "The more you make a tool or program, the more a glitch could have a much larger consequence than it could in the past. You have to have more robust test frameworks or air testing. You need much more robust safety cushions.

It's important to build more risk mitigation processes. If you have no unit testing, if you had only a month to test every permutation, you won't have that agility."

Another thing to consider, says Valadares, is the importance of adopting all aspects of DevOps culture, as identified by Chris Kelly, developer/evangelist at New Relic[16]:

- Hire generalists
- Integrate continuously
- Monitor continuously
- Deploy continuously
- Build with resilience

"You can't take one of the aspects of the culture, you have use all of them or it will be very risky," says Valadares. You have a working methodology like Agile. Then you have the necessary tools. You must have a cultural expectation like Fail Fast. Then you need automation." If all of these are working together, you then have the optimum environment for DevOps to succeed.

---

[16] Kelly, Chris. 5 Keys to Developing a Successful DevOps Culture.

Finally, hiring people experienced in DevOps is a crucial, yet difficult task. It's much more challenging to staff for DevOps." says Valadares. "You need to have the classic system admin skill set, but the DevOps person also needs more developer skills and development best practices. Ultimately you need someone who understands both. You're looking for developers who like back end or sys admins who like code. Then you have a devops."

The human challenge and benefits of DevOps

It is clear that high-performing IT departments can create a competitive advantage for their companies. And as we consider the CIO shaping business strategy to help their companies succeed in the modern business environment, it becomes crucial to remove the blockers that could impede high performance. A change to DevOps may be the way to help people develop more efficiently, as API-led connectivity is an IT operations architecture that increases efficiency.

The current interest in DevOps is a result of the evolution of tools and culture. "Agile methodology—e.g., scrum or combine—have also helped organize the flow of work," says Valadares. "10 years ago people were working in parallel, now they are working collaboratively. All of these things—culture. tools. methodology—have been evolving. And people are evolving as well as products and platforms. You can't expect to transform your business on a waterfall approach. Everyone is headed towards the iterative model, which has reached a maturity point."

There are many ways to organize IT architecture and ultimately you have to choose the one that works for your company. What is intriguing about DevOps is that it uses similar principles of API-led connectivity - composability, empowerment, and reusability- to improve the effectiveness of people as well as what they are working on. A bottom-up approach and the power to create new things is an effective strategy no matter where it's applied in the business.

Part III

Customer stories

It's time to move beyond the realm of the theoretical into the practical. Here are three vignettes about companies whose IT decision makers decided to change the way things have always been done and reimagine how their business uses technology. Each of these CIOs or CTOs used the power of API-led connectivity to achieve business goals: speeding up operations, centering their business around the customer, and creating new revenue streams. They've used the principles of composable, reusable services to create new experiences for their audiences, and as a result, they've enjoyed tangible benefits.

CHAPTER 16

News Corporation: The need for speed

by Ahyoung An, MuleSoft

News Corporation is one of the largest media companies in the world and the largest in Australia. It's an $8.57 billion company and owns major newspaper titles like the *Wall Street Journal* and the *Sun*. In Australia, it owns 150 titles and touches 75% of the Australian population over the age of 14 every week.

The risk of losing consumers and advertisers

News Corporation found itself profoundly affected by digital disruption in the news business. Mobile web page views rose six times in two years and mastheads were facing double digit declines year-on-year. Consumer behavior was changing rapidly and News Corp had to make a choice: adapt quickly and transform their business or risk losing everything.

Transformation meant maximizing their greatest asset — their content — by delivering them through new channels and methods consumers would pay for, while also creating engaging environments for advertisers to connect with consumers. As News Corp Australia CIO Tom Quinn says, "If we can't deliver content quickly when the customer wants it, at 6 AM on their smartphone or whenever they need it, then we'll go out of business."

To execute this vision, they needed to quickly produce, test and market new ways of delivering content through multiple channels to everyone they could reach. Ultimately, speed and agility would determine their ability to compete and win. Quinn notes, "I have a saying with my staff—speed isn't a big KPI, it's the only KPI."

Fulfilling a radical vision with a transformed approach

Recognizing the urgency to transform, a radical cloud-first initiative was launched by News Corp, with Tom Quinn as the driving force. News Corp Australia set the goal of moving 75 percent of the existing enterprise applications to the cloud by the end of the year with all new solutions being cloud-only. This would allow them to remain nimble, move fast, and make changes rapidly. News Corp also wanted to evolve their API strategy by emphasizing composability and reusability - again, to increase speed and flexibility.

Their current connectivity solution was not feasible as it wasn't cloud-based and it didn't let them move as quickly as they wanted. They embarked on a search for partner who could provide a simple, scalable cloud platform and help them evolve their API strategy and their approach to connectivity.

With MuleSoft, rather than creating bespoke, individualized APIs, they decided to create purpose-driven, modular APIs that can be reused. This resulted in a new suite of APIs that would take content from their repository and enable text, photos, videos, and links to automatically populate in their new tablet application with minimal work. With this new model of modularizing their content delivery process with APIs, they no longer needed to undertake manual work previously required to manage the process.

3x faster and higher customer retention

Since adopting an API-led approach to connectivity, News Corp is moving three times faster, has reversed decline in readership and is growing pageviews and customer retention. With the ability to distribute content wherever they want - websites, tablet apps, mobile devices - quickly and efficiently, customers get the content they need when and where they need it, and advertising rates are back on the rise.

News Corp has already exceeded their cloud-first goal with 76 percent of their enterprise apps migrated to the cloud. Integrations that were taking them months are now taking days, allowing them to easily assemble various SaaS components to bring new products to market quickly.

According to Quinn, "We're very positive that we can use these technologies in the right way to continue to grow, and MuleSoft will be the key critical component of what we do. The future of technology at News Corporation is Anypoint Platform."

CHAPTER 17

Top UK retailer creates better customer experiences with APIs

by Sarah Burke, MuleSoft

One of the top 4 grocery stores in the UK was facing numerous challenges in the changing retail environment. Operating in a competitive retail market, the company needed innovative ways to drive in-store revenue while also expanding into new product categories and channels on the web. In addition, the retailer recently acquired an online baby goods store, and needed to understand customer behavior to provide customized offers and promotions both in-store and online. This retailer's Middleware Architecture and Design Manager had the responsibility for bringing existing customer and product data into the company's new Salesforce solution, providing a single view of the customer.

Challenge - how to aggregate customer data when it's locked in legacy systems

"The fastest growing grocery segment is online, and to continue to grow our business we needed to expand to new channels on the web while also reinvesting in our core in-store business," said the middleware manager. The grocery market is experiencing fierce competition for market share. The company needed to compete aggressively in this existing market while also extending their grocery business to the web where sales are expected to double by 2018.

After acquiring an online baby retailer, the company had a high performing online channel along with new customers in the higher margin baby product category. To capitalize on this new channel and customer base, the company expanded the baby brand to a new series of retail superstores while looking to integrate these customers into the brand experience at their existing grocery stores.

However, customer and product data were stored in multiple databases, an IBM Unica marketing automation platform, and an Avaya call center platform. The middleware manager and his team made the strategic decision to consolidate a single view of the customer in Salesforce to run multi-channel marketing programs and promotions across brands. "Our focus is on delivering value to our customers and we wanted the ability to extend our relationship with the customer so we could offer insightful promotions and recommendations tailored to the needs of each individual."

In looking for a solution the team concluded early on that they needed a real-time integration platform. According to the middleware manager, "Since we had multiple endpoints handling a high volume of transactions requiring real-time synchronization, a solution with an ESB approach made more sense to us than a batch ETL tool."

They were also interested in accomplishing the integration using the same cloud-based approach they took with the marketing initiative that led them to Salesforce because of its lower operating costs and ease of use.

Solution - a cloud based integration platform providing visibility into customer data

The company worked closely with their implementation partner to develop the integration solution. After a head-to-head comparison, they selected Anypoint Platform as their single platform for application integration. Developing and deploying the integration on CloudHub, the managed cloud platform for Anypoint Platform, took just 2 weeks since the platform requires no hardware or software. "One of the important criteria for this project was a 100% cloud solution, and CloudHub was the only solution that delivered enterprise-class capabilities without agents to install on premises," said the middleware manager.

During the initial rollout they connected Salesforce to their on-premise call center platform, and connected Salesforce to customer and product data stored in on-premise databases using Virtual Private Cloud.

The company's call center handles inbound calls from the baby retailer's website. Since customers are calling to troubleshoot orders online or order directly over the phone, having complete visibility into all information related to a customer enables representatives to increase sales.When a customer calls and hangs up before reaching an agent, this information is captured in Salesforce and representatives can call the customer back, providing a much higher level of customer service.

Benefits to an API-led connectivity approach

With Anypoint Platform, the company achieved a single view of the customer in Salesforce in just 2 weeks. With no hardware or software to deploy and no firewall policy changes or proxy servers required, integrations were fast to configure. The company can now begin multi-channel marketing, armed with data that makes their messaging timely and relevant. This initiative is expected to be a key driver of growth and competitive advantage in the future. With its multi-tenant architecture, today Anypoint Platform supports real-time synchronization of 5 million customer records and hundreds of thousands of SKUs.

As the business grows, the company is confident that Anypoint Platform will scale to support even larger volumes.

Looking to the future, the company plans to build their digital marketing strategy on Anypoint Platform, creating intelligent campaigns based on customer behavior. "We wanted more than an integration tool. MuleSoft's product is the platform to build our business as we expand grocery sales online and develop additional marketing promotions across channels to grow our business,"said the middleware manager.

CHAPTER 18

Addison Lee: Reinventing business with APIs

By Aaron Landgraf and Shana Pearlman, MuleSoft

Addison Lee is Europe's largest premium car service, operating in more than 350 cities and transporting 10 million passengers every year. It was the first car service to utilize a mobile app and SMS messaging, which it released in 2009, pioneered fleet allocation software to match drivers with customers which is now sold to and used by fleets around the globe, and today has a strong digital focus on engaging with its customers, recently being one of the first in the UK to launch Apple Pay and an app for the Apple Watch. "We now take over 50% of bookings by mobile, and that has been led by customer demand," says Chief Technology Officer Peter Ingram. Keeping at the leading edge of customer service and technology is crucial to Addison Lee as the company faces competitive pressure from new technology focussed entrants in the marketplace.

Changing customer behavior required faster delivery

Addison Lee has always been at the forefront of the taxi and mini-cab service leveraging the latest technologies to deliver best-in-class customer satisfaction. "Customer behavior is changing. They want to pay by credit card. They want it to be very easy to use. They are not necessarily wishing to phone up and book, but they like the assurance there is someone there if they need to talk," says Ingram. Since Addison Lee has always been disrupting the ground transportation industry, before disruption was a business buzzword, they decided to create consumer-friendly innovations on a regular basis to keep up with customer demand. Software development needed to happen faster, and couldn't consist of bespoke code anymore - it needed to be composable and repeatable.

Unlocking data and infrastructure with APIs quickly

In just 6 weeks, Addison Lee securely unlocked their data and infrastructure with their first public API built on MuleSoft's Anypoint Platform. This was specifically designed to support development of booking apps and web sites by third party affiliates and partners to incorporate the Addison Lee service. "We have a team of around 50 developers working full time on a number of our enterprise platforms, including Salesforce as well as our own bespoke allocation software, and support for digital development. The MuleSoft technology allows us to rethink how we connect our systems and expose our data and services in new ways to support a creative mobile strategy," says Paul McCabe, Head of Addison Lee Development.

One important new revenue stream for Addison Lee is being able to seamlessly connect its mobile application users with its existing network of international fleets. Soon a customer will be able to book a car in a large number of global locations using the Addison Lee app instead of doing this through the contact center. "MuleSoft is enabling us to look at new standards for connecting to partner car fleets globally," McCabe says.

MuleSoft reduced integration times from weeks to days

Addison Lee is fully on board with the composable enterprise philosophy, pulling together numerous back-end systems, plugins, and third-party offerings though API-led connectivity. To make this manageable, MuleSoft has enabled Addison Lee to dramatically reduce the time it takes to integrate back-end systems, plugins, and other services. CTO Peter Ingram says, "Now that we have MuleSoft in place, we are starting to see we can reduce integration times from weeks into days. That will make us much more nimble as we grow our integrations. We can offer new products and services very quickly." In the past we would typically take a bespoke approach to integration, but today our focus is on APIs and management through MuleSoft."

MuleSoft will be the backbone for global integrations, for a new Service Oriented Architecture and customer innovation at Addison Lee. "API-led connectivity with MuleSoft is key to our growth aspirations," says Peter Ingram. "It empowers us to be creative in how we connect our systems to deliver new digital products and offers, and to support new third party integrations, both with our customers and our partner network of fleets."

Part IV

Your integration assessment

CHAPTER 19

Your integration assessment

by Sarvesh Jagannivas, MuleSoft

By now you've seen in detail the principles of the composable enterprise and how they can create success in a hyper-competitive business environment. Perhaps you are on the road to enacting some of these changes. But how ready is your business to make change? Using the principles of the composable enterprise, the table below provides a broad framework for assessing and measuring the success of IT departments in the transformation into a new innovation center for the company.

Parameter	Context	Assessment guidance
Reusability of components	When you build more reusable components, teams are faster on subsequent projects. Examples of reusable components include APIs, services with canonical representation for commonly used data records such as employee, financial, customer, etc, standard connectors, integration patterns depicted as templates.	What is the percentage of new applications that use previously built components? What is the percentage of new applications that build components that others can access and reuse?

Parameter	Context	Assessment guidance
Ease of use	Design, build and test: Teams are faster in design, development, and test with easier tools. Examples include two-way editing, embedded test tools, collaboration tools for multiple designers to work on concurrently, off the shelf connectors to popular applications and standards.	Does the technology require proprietary tools - percentage of open technology in your tooling? Do you have to go into code for doing anything (percentage)? What is the training time needed for on-ramping new developers?
	Operate: With unified monitoring across all deployments and with a secure, elastic, multi-tenanted, resilient and fully managed Cloud, running and operating applications in production is more productive. Less time is spent in keeping the applications running.	Do you have common monitoring and management across all deployments (hybrid)? Do you have a secure, fully elastic, resilient, multi- tenanted, fully managed Cloud or on-premises deployment?

Parameter	Context	Assessment guidance
Ease of use	Diagnostics and test: Ability to diagnose applications and APIs quickly, isolate bugs, fix them and conduct regression tests. This helps teams run faster and keep error rates low.	What is the average time taken to make changes or fixes?
Self-serve capability	Downstream app developers and line of business developers depend on Central IT for access to data. The more these developers can self serve, the faster they will go, while Central IT maintains governance and control.	What is the average time taken to find existing assets such as APIs, documentation to understand them, and get ready to use them to construct further APIs and applications?
Built for purpose	Most often, enterprises boil the ocean, in which a whole number of services and APIs are built for future use, without necessarily having an end application in mind. If APIs are designed and built for a specific immediate application need, team productivity goes up because they are working on the most timely projects.	Do you start with the end user and end use defined before you start designing the API? Do you collaborate with the end user of the API or applications development team?

It's a given that the nature of doing business is changing. The choice to do nothing is a very risky one. Connectivity is at the center of business transformation, enabling organizations to adapt, innovate and transform faster than the competition. Now there is an established roadmap for enterprise technology success.